The Rat

The Rat

A Study in Behavior

S.A. Barnett

 ALDINETRANSACTION
A Division of Transaction Publishers
New Brunswick (U.S.A.) and London (U.K.)

Library of Congress Catalog Number: 2006052670
ISBN: 978-0-202-30977-4
Printed in the United States of America

Library of Congress Cataloging-in-Publication Data

Barnett, S. A. (Samuel Anthony), 1915-
    The rat : a study in behavior / S. A. Barnett. / Lyn H. Lofland.
        p. cm.
    Originally published: Chicago : University of Chicago Press, 1975.
    Includes bibliographical references and index.
    ISBN 978-0-202-30977-4 (alk. paper)
    1. Rats—Psychology. 2. Rats—Behavior. I. Title.

QL737.R666B37   2007
599.35' 215—dc22   2006052670

Contents

Illustrations

PLATES

TEXT FIGURES

Preface

It is to be expected that some research workers will always like to watch animals, while others will prefer cathode ray oscillographs; and that, of those who study behaviour, some will crouch over rats in problem boxes while others lie in bogs peering through a binocular. This is an inevitable, and welcome, consequence of human polymorphism. In this book it is assumed that, within the framework of scientific endeavour, all these activities are desirable and that none is morally superior to any other.

My intention has been primarily to present some of the principles of ethology, that is, of the study of behaviour. The concentration of behaviour research on laboratory rats has made it possible to take most of the examples from one species. Something is now known, too, of the behaviour of wild rats; hence observations and concepts, of kinds usually drawn from the work of zoologists, can easily be introduced with those of psychologists; I have indeed laid special emphasis on biological principles. Each group of workers has much to contribute to the work of the other. It was, however, not enough to describe observations only of overt behaviour. An animal's behaviour is not something apart from its physiology but is a reflexion of the working of all its organs. The study of behaviour makes demands, and has rules, of its own; but in one aspect it is a branch of physiology. Accordingly, I have described studies of the nervous system and the endocrine and other organs, wherever they help to explain behaviour.

I have assumed that communication about behaviour is aided by rigorous definition of important technical terms and by holding firmly to each definition once it has been given. The definition of 'definition' is discussed in an appendix. No doubt I have often failed to keep to my own principles; if so, I hope at least that others will be provoked into doing better. A disadvantage of being clear is that one must sometimes be clearly wrong. But it is better to risk that than to resort to obscurity.

The attempt to be lucid has obliged me sometimes to criticize opinions no longer held by their original authors or by present authority. This is only because obsolete theories and notions often reverberate in teaching when their usefulness has long been exhausted. One example is found in conventional accounts of the conditional reflex: whole

xv

classes of writers, for instance of textbooks of physiology, have persisted for decades in obvious error on this subject. I have also tried to throw light in areas where certain writers have fallen with a crash into logical traps and, once fallen, have wallowed in them, to the confusion of their pupils. None of my criticisms is directed *ad hominem:* indeed, I owe much (as will easily be seen) to the people criticized.

I have tried to give clear acknowledgement wherever I have quoted the observations or ideas of others, except where recent comprehensive reviews have been available as secondary sources. For most chapters or sub-chapters there are general references in the first section of the bibliography. These works are usually mentioned also in the text. In the passages in which only general references are given (though the names of authors of particular items of research may occur) the original sources must be sought in the reviews. But in most parts of the book each account of particular researches is preceded by an author's name and ends with a number in square brackets; the latter then refers to an entry in the main bibliography. Occasionally, the reference is given in the caption of an accompanying text figure.

The great volume of work on rat behaviour (let alone that of other animals on which I have had to draw) has imposed difficult decisions on what to quote. Much good work has been left out. Where I could I have given preference to clear and well-organized writing, but this has usually not been possible. The advance of ethology (as of other subjects) would be greatly helped if the standard of writing about it could be raised. There are some outstanding examples for imitation: for instance, the late K. S. Lashley from American psychology and N. Tinbergen from European zoology.

Writings on scientific subjects have, as a rule, only a brief usefulness at best. This book will certainly not be an exception. If they are sufficiently lucid they can contribute for a little to the continuous disputation which must accompany both research and teaching. The most valuable things I can hope to have done are to heighten interest in the science of behaviour, to provoke argument about it and, finally, to encourage readers to adopt the attitude of the motto of the oldest scientific society: *Nullius in verba.*

S. A. B.

THE RAT: A STUDY IN BEHAVIOUR

I

Statement of Themes

'Tis laughable that man should fondle such surprise
at animal behaviour, seeing some beetle or fly
– whose very existence is so negligible and brief –
act more intelligently than he might himself
had he been there to advise with all his pros and cons,
his cause, effect and means: Such conduct he wil style
'Marvels of Instinct', but what sort of wisdom is this
that mistaketh the exception for the general rule
and the rule for the exception?

ROBERT BRIDGES

1.1 NATURAL HISTORY

A much-esteemed experimental psychologist once published a denunciation of the excessive use of tame rats in laboratory studies of behaviour. His article, inspired by a work of Lewis Carroll, was entitled 'The Snark was a Boojum'; it claimed that comparative psychology had softly and suddenly vanished away under predation by *Rattus norvegicus* var. *albinus* [38].

1. *Rattus norvegicus.* An adult male on the run.

It is true that *comparative* ethology cannot exist without information on the behaviour of many species; but it is still possible to illustrate many of the principles of ethology from one genus, or even one species, only. Accordingly, the species with which this book is mainly concerned

is *Rattus norvegicus* Berkenhout, the common 'brown' rat of Europe,
North America and elsewhere. It is sometimes called the 'Norway rat'.
Albino and other tame varieties of this species have been used in laboratories for more than half a century. Another species which will be
mentioned is *Rattus rattus* L. This, variously called the black, house,
roof, ship or alexandrine rat, is less common in Europe and North
America. It varies greatly in colour. Individuals of both species can be
tamed, if they are regularly handled from an early age [26]; but there
are no laboratory strains of *R. rattus*. *R. norvegicus* is a burrowing
animal (figure 2) but it can climb well; *rattus* does not burrow, but
climbs even better and typically nests above ground.

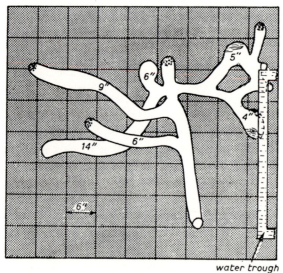

2. System of earth burrows of *Rattus norvegicus*. Figures show depths. (After
Pisano & Storer [244].)

Both species have a long history of association with man, well described by Barrett-Hamilton & Hinton [34]. It is possible that living
in human communities, and competing with man for food, have led, by a
process of natural selection, to genetical changes influencing behaviour;
this applies especially to the extreme fear or avoidance of unfamiliar
objects (such as traps) displayed by wild *norvegicus*. The omnivory
which rats display may, however, have evolved before human communities become prosperous enough to support populations of small rodents.
   Most studies of rat behaviour have been on tame varieties of *norvegicus*. These laboratory rats can easily be crossed with wild ones, but

their behaviour differs from that of wild rats in many important respects, not all of them obvious. They have been selected for tameness, that is for a reduced tendency to flee from man or to fight when handled. In addition, strange objects which would induce 'fear' in wild rats provoke only 'curiosity' in tame ones; and tame males do not fight other males with anything like the same intensity as wild males. Their selection of foods, when faced with a choice, may also differ from that of wild rats. A *similarity* of behaviour, which they are rarely allowed to display, is that they burrow readily.

Differences in behaviour have been paralleled by changes in growth, many of which have been described in a standard work by Donaldson [87]. There are also important differences in the relative weights of organs, such as the adrenal glands. A population of wild rats is of course genetically heterogeneous, and it will be shown later that generalizations about the behaviour of wild rats often have to be qualified by reference to individual variation. In an inbred strain of tame rats genetical variation is greatly reduced. This has the advantage, for certain kinds of experiment, that differences between experimental and control groups are more likely to be the effects of the experimental procedure, and less likely to be due to genotypic variation with which the experimenter is usually not concerned. Nevertheless, it is misleading to speak of 'the laboratory rat' in a way which might imply that laboratory rats are quite uniform. First, some genetical variation always remains. Second, in inbred strains, minor differences of environment can produce marked phenotypic variation: in this respect, highly inbred animals may be less convenient material for experiments than the first-generation hybrids between inbred strains; such hybrids are both genetically *and phenotypically* often remarkably uniform. Their use may make possible a reduction in the number of experimental animals needed to give conclusive results [214]. Third, and above all, there are several varieties of tame rats in common use: in addition to the familiar albinos, which have pink eyes and poor eyesight, there are coloured types – hooded, piebald, grey and black – which have a pigmented iris and correspondingly better vision [140].

The existence of a number of tame strains of a species which is also easily available in the wild form makes *Rattus norvegicus* a particularly convenient animal in which to study behaviour. This convenience has its dangers: a concentration of research on one species could lead to the production of an absurdly incomplete and distorted picture of animal, or at least of mammalian, behaviour. Accordingly, in this book, mention is made of other species wherever it seems important to emphasize resemblances to rats or differences from them; or when some aspect of behaviour has been insufficiently studied in rats. However, of all mammals, rats are those which, with present knowledge, can best be used to illustrate the principles of the study of behaviour. The rest of this chapter introduces the main definitions, and general concepts, of which use has to be made in such a study.

1.2 THE DEFINITION OF BEHAVIOUR

By behaviour is meant here the whole of the activities of an animal's effector organs: in the case of a mammal and most other sorts of animal, these are its muscles and glands. This definition includes the contraction of smooth muscle and the secretions of all glands, and is intended to do so. Nevertheless, most of this book is concerned with behaviour in a rather narrower sense, namely, the movements of the whole animal — movements which depend on the activity of many skeletal muscles. This does not necessarily exclude facts which are commonly thought of as coming within the realm of physiology. One way of studying gross, overt behaviour is to seek the internal processes which influence it, especially those in the sense organs and the nervous system; but this in turn demands enquiry into such things as the effects of hormones, the level of sugar in the blood and so on. Further, the study of the senses should not be confined to those that convey information about the world outside the animal (the exteroceptors), but should also comprehend the internal sense organs: of these, the proprioceptors, which are present in the organs in which most movement occurs (the muscles, tendons and joints), play an essential part in every act of behaviour.

The scientific study of behaviour (ethology) is therefore not marked off sharply from physiology, of which indeed it could be regarded as a branch. Nevertheless, ethology has methods of its own, different enough from those used in laboratories, or described in texts, of physiology for it to deserve a separate name. This is partly because a physiologist is usually concerned with units of activity smaller than those studied by ethologists: research on changes in salivation rates due to stimulation of the chorda tympani nerve would usually be done by a physiologist; but an inquiry into salivation in response to the smell of food would be more likely to be done in a department of zoology or psychology. In the second case the experimenter is observing the operations of a unit which includes the olfactory organs and parts of the brain, as well as the motor nerve fibres which supply the salivary glands and the glands themselves.

The difference in size of unit exists also, as already implied, in the responses studied. A pattern of muscular contraction, as in a tendon reflex, is the type of 'molecular' response which is within the domain of physiology. 'Molar' responses, involving much more complex relation-ships, belong to the study of behaviour as it is usually practised. These larger responses bring about either a change in the environment (for example by moving some object, or dissolving it in a glandular secretion), or they alter the relation of the animal to its environment (for example, by moving the animal off to another place).

Whatever form is taken by the activity studied, the behaviour is always a combined result of the actions of many organs. In elementary texts this fact is sometimes represented by a diagram of a 'reflex arc', in which a single receptor (such as a sensory organ in the skin) is shown connected to a single afferent nerve fibre, and this in turn to a single connecting neuron in the central nervous system; the arc is completed by one motor neuron ending on one muscle fibre or gland cell. This is useful in so far as it conveys that overt behaviour is a result of processes in the organs symbolized in the diagram. But even a 'simple' reflex is a result of a complex interaction of units in the central nervous system: there is no one-to-one relation between sensory, connecting and motor nerve cells; and several muscles are involved, some contracting, others relaxing, all in an orderly way. Further, some neural processes are *inhibitory*: they are not, as many diagrams imply, all excitatory. In a simple reflex there is progressive relaxation of the muscles antagonistic to the contracting ones; and this depends on inhibition in the central nervous system. A 'simple' reflex as a whole is a highly integrated performance depending on the smooth functioning of large numbers of nerve cells.

More complex activities involve correspondingly greater elaboration in the central nervous system. This applies to most many-celled animals, but above all to the mammals, in which a vast mass of connect-ing neurons is interposed between receptors and effectors. This mass of nerve tissue does not function in the machine-like way suggested by the reflex arc diagram, and not very much is known of how it does work. Nevertheless, despite our ignorance of brain function, it is still helpful to speak in terms of the control of behaviour by the nervous system, since otherwise we are tempted to talk as though animals were controlled by a set of gremlins—urges, drives, instincts and so on – and this does not help us to understand behaviour as it actually is. At the same time, we have to use words which refer to the different kinds of overt behaviour – the activities that we can observe without regard to internal processes. It is to these activities that we must now turn.

1.3 'LEARNING' AND 'INSTINCT'

When an animal performs some action, it is often said to have done so in response to a stimulus. The term stimulus may be formally defined as any event which causes a change in an animal's receptor organs and, as a result, in its nervous system. This definition is very general, and lacking in precision since it leaves open the meaning of 'receptor organ'; but in practice it is convenient, and it does not encourage ambiguity. Sense organs may be looked on as instruments which measure either the intensity or the amount of some form of energy falling on them. To make sense, any statement concerning stimuli must specify (or imply) what kind of energy (light, pressure and so on) is involved. The definition of 'stimulus' excludes changes in the surroundings to which the animal's sense organs cannot respond: for instance, a change of colour in an object cannot by itself influence a rat's behaviour, because rats have no colour sense.

The sense organs and the nervous system are so arranged that the impact of a very small charge of energy can release a much larger amount of the energy stored in the animal's muscles. If a rat is sitting still, and a short, high-pitched sound is made near it, the energy of the sound waves falling on its eardrums is trivial; yet the animal may leap to its feet and run to cover.

The nervous system not only has this amplifying effect: it also determines the pattern of the response. The rat disturbed by a sound runs by a direct and economical route to cover, provided that it has already had the opportunity to move about the area on previous occasions. But if it is in a strange place, although it will run wildly about, its movements will not be related to the whereabouts of shelter. This is an example of the kind of change that often takes place in an animal when it repeatedly encounters the same circumstances; we know of this change through altered behaviour. The central nervous system is the organ which, by very complex means, makes this alteration possible.

An important feature of such changes of behaviour is that they are adaptive: that is, they tend to make more probable the survival of the individual or the species. On the one hand the rat running to cover may escape a predator. On the other, a female may respond to the squeaking of her young by returning to her nest and allowing them to feed; this may not affect her own chance of survival, but it increases that of the young rats and so, in general, of the species. In both examples the animal has learnt the route to be followed.

Altered behaviour of this sort is an example of individual or 'physiological' adaptation. It must be distinguished from genetical, or evolutionary, adaptation, in which the genotypic make-up of a population is changed. Physiological adaptation occurs in all animals. An organ which is subject to much use, instead of wearing out like part of a machine, often enlarges or becomes more efficient in some other way. The type of physiological adaptation that goes on in the central nervous system is exceedingly complex: it is sometimes called 'learning'. Learning in this sense has been defined as internal processes which manifest themselves as adaptive change in an individual as a result of experience [311]. However, the variety of learned behaviour is very great, and it may be doubted whether convenience is best served by grouping all the underlying processes under one term.

Another, still more dubious grouping is that which divides complex behaviour into two kinds, the innate (inborn or, sometimes, instinctive) and the learned (which includes 'intelligent'). This distinction must now be examined.

In the behaviour of any species there are certain patterns of movement which are readily recognized as 'the same' when performed by any individual. (Some are confined to one sex.) For instance, the behaviour of a male rat approaching and copulating with a female in oestrus consists of a fixed sequence of actions. The same applies to the behaviour of the female in oestrus when approached by a male. A pattern is a *repeated* set of relationships, and this is found in these examples. These activities are sometimes called fixed action patterns; in this book, when they have to be referred to in general, they will be called stereotyped behaviour. They are, to repeat, typical of the species (or other taxonomic group) to which the animal belongs and so are often said to be species-characteristic.

A fixed action pattern is often evoked with a high degree of predictability by a specific stimulus. Such releasing stimuli have been studied in great detail in some species of birds and fish, but not so fully in mammals. In the case of the male rat and the female in oestrus, the mating behaviour of the male is evidently released, in the most usual situation, by the combination of a characteristic odour with the visual pattern made up of the shape and movements of the female; once the female has been mounted there are also tactile effects. The fact that a highly specific pattern of behaviour (that is, of movements) is released by an equally specific stimulus pattern (however complicated), gives a machine-like impression to the observer: one is tempted to speak, by analogy, in terms of such things as triggers which, when pressed, produce automatically a predetermined system of movements. The internal 'mechanism' involved is obviously the whole organization of sense organs, nervous system and muscles, but there is evidently a special arrangement in the central nervous system which determines the precise pattern of actions involved.

So far, then, we can say that the behaviour of an animal in response to a particular stimulus or situation may obviously have been modified, in which case we say it is a product of learning; or it may be stereotyped, in which case we often do not know to what extent learning has played a part in its development. Whatever the degree to which behaviour is stereotyped, we can usually see that it is related to the biological needs of the animal: it has 'survival value'. The relationship of behaviour to need must now be examined further.

Fixed action patterns sometimes come as the end of longer sequences of activities. A rat feeds after it has moved from its sleeping place to food; the movement may be direct, if the food is in a place in which it has been encountered before, or it may be wide-ranging. We have seen that the same difference may be observed in the behaviour of a rat, sitting in the open, when it is disturbed: it may or may not be able to run straight to cover. In the case of the rat moving to food, the actual feeding is an example of a fixed action pattern but the preliminary movement, which may be over a wide area, is of a different character. The difference does not lie in the locomotory movements themselves, since these are stereotyped: one rat walks, climbs, runs or jumps in much the same way as another. The difference is in the relation of the movements to the surroundings of the animal. When a male rat mates with a female his activities necessarily have a particular relationship with her; when he is eating, he is similarly orientated on the food; but when he is moving around, as a *preliminary* either to mating or to eating, there is no standard pattern of movement with regard to the objects about him. The movements are therefore difficult to predict.

The variable movements of animals are often attributable to an internal change corresponding to some biological need, for instance for food. They have consequently sometimes been called 'appetitive behaviour'. Although such behaviour is not standardized, it may become fixed for a given individual. If a rat settles in an area in which it has a nest in one place, and there is a constant source of food in another, it may move regularly from one to the other. Such regularity is a product of the past experience of the individual: it has been learnt. In general, the variable phase of behaviour is the one in which the effects of learning are most noticeable. As a result, at first, of relatively undirected movement, an animal reaches a place where some stereotyped act can be performed. There may then be a period of quiescence. We therefore have the following sequence. (i) An internal change takes place: for instance, hunger develops. (ii) Preliminary, unstandardized movements are performed; this may be called the appetitive phase and is the occasion for learning to take place. (iii) A stereotyped act is carried out, as a result of which the internal condition which initiated the whole process is reversed; this is called a *consummatory act.* (iv) The animal becomes quiet or begins a different sequence of acts.

We can now see more clearly the connexion of these different phases of behaviour with need. The behaviour sequence is initiated by the state of need, and it ends when the need is satisfied. The state of satiation is called a *consummatory state*: this may (for example) be a particular skin temperature, achieved by building a nest or moving to a warmer place; or it may be a full belly or the presence of a substance such as glucose in a particular concentration in the blood.

This simple scheme will be more critically examined in later chapters: it has many imperfections, but it provides a convenient frame of reference for descriptions of behaviour.

1.4 EXPLAINING BEHAVIOUR

1.4.1 *Subjective and objective*

The preceding outline has been wholly in terms of overt behaviour which can be directly observed in the intact animal, and of physiology which can be studied by well-established laboratory methods. No reference has been made to the feelings of rats, their thoughts, their minds or any of a number of such concepts normally used in speaking of human behaviour. This restriction is usual (though not universal) in scientific communication, and we may now inquire whether it is really necessary. Certainly, most people seeing a rat sniffing around in the neighbourhood of food find it both convenient and sensible to say that the rat is hungry and looking for food; and, if the animal is then disturbed and runs away, to say that it has been frightened.

It is indeed difficult to speak objectively about behaviour because a human being ordinarily describes the things he sees by reference to other, more familiar things; and the most familiar behaviour is one's own. In the attempt to explain behaviour, we attribute our own awareness, feelings and thoughts, not only to other people (as, with due caution, we must), but also to other species. This anthropomorphism can lead to error in quite simple ways. For example, like most mammals, rats are greatly influenced by odours. A male can distinguish the odour of a female in oestrus from that of a non-receptive female. An unwary human observer, incapable of this olfactory *tour de force*, might attribute the movement of a male towards a female out of sight or hearing to some mysterious and indefinable agency – called an 'instinct', perhaps – beyond ordinary understanding. Rats also make and hear sounds of too high pitch for our ears. These are examples of the different sensory abilities of different species: the 'world' of sense impressions of an animal much influenced by smell, and responsive to sounds of high pitch, is quite different from that of one relying primarily on vision.

Another kind of example involves the attribution of a high degree of 'intelligence' to rats. Wild rats are difficult to kill because they avoid traps and poison baits. This behaviour often appears intelligent, and it has led at least one biologist to call the conflict between rats and men 'a veritable battle of wits'. However, careful observation of rats in simple experimental situations has shown that the avoidance of strange things is quite indiscriminate: it extends to harmless objects and even to wholesome food; it is 'automatic' rather than intelligent in any sense. Of course, it greatly benefits the animals that display it.

Other examples of the dangers of anthropomorphism will be found later in this book. However, comparisons with human behaviour also have their uses. Indeed, our tendency to *compare* the behaviour of other species with our own is not only 'natural': it is inevitable. It leads to error when it is assumed, perhaps unconsciously, that other species resemble us in matters in which, in fact, they differ from us profoundly. Comparisons must be made; but the *assumption of similarity* must be avoided.

Whatever method of speaking about behaviour we use, the first object, in a scientific study, is to describe it accurately and in some detail. This is far from easy. Secondly, we have to explain it. The two processes are in practice not separate.

In this book, two main sorts of explanation are used. First, any feature – structural, chemical or behavioural – of any species, may be interpreted in terms of its *survival value*. For instance, an adult, male, wild rat in its territory attacks strange males of its own species. If one asks why it does so, the answer may be that the process of natural selection has favoured a genotype which brings about this type of behaviour. It may be said, or hypothesized, that such territorial behaviour has a favourable effect on the structure of rat populations, and so confers a selective advantage on those that display it; this advantage could be due to the dispersal of rats giving access to sources of food. The 'causes' of the behaviour, implied in this statement, are events in the comparatively remote past which have acted in the first place on the ancestors of the animals observed.

Secondly, the behaviour of the male rat may be explained in terms of events which influence it more directly, either at the time of the activity in question or at least during its own life history. This kind of explanation has several sub-divisions. (i) Behaviour is influenced by external stimuli. These may (a) produce an effect at once, as when the sight and odour of another male releases attack; or (b) the past experience of successful fighting with other males may help to induce the male to attack again: this is an example of the effects of external stimuli which have acted earlier in the life of the animal. (ii) Behaviour is also influenced by internal states, such as the amount of male hormone in the blood; a young male does not fight until its testes begin to secrete this hormone in relatively large amounts. Of these two kinds of explanation, type (i) may be said to be behavioural, type (ii) physiological. None of these sorts of explanation is incompatible with any other, nor can one replace another. Often, one method of analysis suggests problems which have to be solved by another method, as when observation of behaviour leads to a study of endocrine organs.

Given the different kinds of explanation, we may ask by what criteria we judge an explanation to be valid or acceptable. There are two tests. A particular observation, or group of observations, may be put in a larger class of phenomena or, more generally, may be incorporated in some accepted frame of reference. If it is asked why male wild rats attack other males, the reply may be that territorial behaviour is very common among the vertebrates, and this is an example of it. Such a statement can be at least moderately satisfying to an inquirer: its main features are that it relates the particular behaviour of male rats to the general proposition that behaviour patterns are usually advantageous (owing to the operation of natural selection); it also relates the fighting of males to the general notion of territorial behaviour.

The second test of an explanation is the more stringent one that it should be possible to base correct *predictions* on it: that is, an explanation is said to be valid when it is a general statement from which it is possible to deduce particulars. The statement that all adult male wild rats attack strange males is of this kind, though it is not one of very high generality; (nor is it quite true). Another example is: aggressive behaviour in male mammals depends on the presence of male hormone in the blood. Both are readily tested by experiment; if the results of experiments are in conformity with the proposition tested, then it is held to be to that extent valid.

There is a further distinction to be made. It is often felt that physiological explanations are in some way superior to others. This is in fact an inappropriate way of expressing a real difference. A physiological account of behaviour has two special features. First, it gives the immediate causes of the behaviour, both external and internal; it should therefore enable us to change the behaviour more easily, more extensively and more certainly than before. Secondly, it relates the internal organization of the animal to its behaviour; this should enable us to predict the way in which the animal will behave from knowledge of its organization. We are, however, a very long way from any substantial achievement on these lines.

1.4.2 *Questions of quantity*

Sometimes, a prediction is correct in only some of the instances in which it is put to the proof: for example, a few adult male wild rats do *not* attack strange males of their own species. It is indeed always true that predictions involve some uncertainty. Often, therefore, they have to be expressed in terms of *probability*. Such statements, based on statistical analysis, are primarily descriptions of the way in which certain parameters are distributed in a population: the figures used for the calculation may be the results of measurements or of counting (for instance, numbers of young per female). As an example, the adrenal glands vary in functional state according to the social relationships in which rats are involved: in one population of males the mean adrenal weight may be 60 mg per 100 g body weight, in another, 80. It is customary and desirable to accompany every such statement of a mean with an estimate of the degree of scatter or dispersion about the mean (such as the standard deviation), or of the reliability of the estimate (usually the standard error). When mean values from two populations are compared, it may be necessary to calculate whether the difference between them is 'real' and not due to 'chance'. This figure, too, is primarily a description of the way in which the two calculated distribution curves are related to each other.

When means or other figures, such as regression coefficients, are compared, the term 'significance' is conventionally used. This term, better rendered as 'statistical significance', has not the meaning it possesses in ordinary speech. When one says that two means are 'significantly different at the five per cent level of confidence', one is still uttering a descriptive phrase. It carries also, however, an element of prediction: it implies that, if one repeats such measurements on many different occasions, on only one occasion in twenty will the observed relationship be reversed. What action one takes as a result depends on a number of factors, most of them not amenable to statistical treatment. If a man, wishing to perform some feat, is told that there is one chance in twenty he will be killed in the attempt, he may decide to forgo the performance. But if a research worker finds a difference, 'significant' in this sense, in the adrenal weights of two rat populations, he may be encouraged to perform many arduous experiments to test, say, the hypothesis that there are corresponding differences in pituitary secretion or hypothalamic function. If he is fortunate, he may find differences of such magnitude that no statistical analysis is needed and a biological relationship is unequivocally established. In that case, everyone, except those who are interested not in biology but only in doing sums, will be satisfied [154].

If some figures are not 'significant' at the minimum conventional level of five per cent, it by no means follows that there is no real difference between the populations compared. Quite often, in such a situation, one may believe, or at least strongly suspect, the existence of a real difference; and this belief may depend on other information, from quite different sources, which cannot be combined for statistical purposes with those one is considering. There is much evidence from studies of laboratory rats and other mammals concerning the state of the adrenal glands in animals subject to stress; this evidence could justifiably influence one's attitude towards observations of adrenal weights.

None of this must be taken as a derogation of statistical analysis, since there are two ways in which it is of the utmost value. First, it is essential for the orderly, concise and rigorous presentation of quantitative material. Secondly, it may reveal relationships in numerical results which are not evident at first. (A graph may do the same in a simple case.) The primary reason why statistical analysis looms so large in biology is that biological material is so variable. Organisms and their parts are in this respect quite different from molecules or atoms. This fact is often concealed by ordinary linguistic usage. One speaks of 'the rat', as in the sub-title of this book, when one is referring to a vast population of creatures, falling into many varieties, of which only a few can be studied. Biologists are fortunate that mathematicians have provided them with an accessory language which enables them to convey precise information about the bewildering diversity they try to understand.

1.5 CONCLUSIONS

To sum up, many of the principles of ethology may be illustrated from a single genus. In describing the behaviour of rats we have to deal with more than one species, and within each species more than one variety. Within *Rattus norvegicus* we have both the wild varieties and also the domesticated ones which are behaviourally very different. Within each variety there is further individual variation, some determined genetically, some environmentally. Inbreeding reduces genetical heterogeneity, but may increase phenotypic variation.

Behaviour is variable for all these reasons; but it is made still more so by the way in which it is controlled. Most behaviour can be described in part as response to stimuli. A stimulus is a small change of energy outside or inside an animal which leads to a large change of energy output by the animal's effectors. The change in energy output, that is, in behaviour, is mediated by the central nervous system. Repeated stimulation of the same kind often leads to changed behaviour: this is due to a kind of alteration of central nervous function which we call learning. As a rule, learning is adaptive, that is, has survival value for the individual or the species. It represents an extra dimension of variability, superimposed on genetical and on other environmental diversity.

Some behaviour consists of species-characteristic fixed action patterns: in them the effects of learning are often small or barely observable. Performance of a fixed pattern is typically preceded by a more variable (less predictable) activity, called appetitive behaviour, during which learning takes place. Appetitive behaviour is often set going by a specific internal state accompanying a need, for example, for water. The activity ends when the animal reaches a goal and is able to achieve a consummatory state.

The arid and polysyllabic vocabulary conventionally used in scientific accounts of behaviour is due to difficulties which arise if one tries to describe animal behaviour in human terms. The senses, abilities and needs of other species differ from ours, often in obscure ways. To convey the maximum of information, students of behaviour have to employ objective and, where possible, quantitative ways of speaking.

2

Movement in the Living Space

The whole story of our dealings with the lower wild animals is the history of
our taking advantage of the way in which they judge of everything by its mere
label, as it were, so as to ensnare or kill them. Nature, in them, has left matters
in this rough way, and made them act *always* in the manner which would be
oftenest right. There are more worms unattached to hooks than impaled upon
them; therefore, on the whole, says Nature to her fishy children, bite at *every*
worm and take your chances. But as her children get higher . . . she reduces the
risks. Since what seems to be the same object may be now a genuine food and
now a bait; since in gregarious species each individual may prove to be either the
friend or the rival, according to the circumstances, of another; since any entirely
unknown object may be fraught with weal or woe, *Nature implants contrary
impulses to act on many classes of things*, and leaves it to slight alterations in the
conditions of the individual case to decide which impulse shall carry the day.
Thus greediness and suspicion, curiosity and timidity, coyness and desire . . .
seem to shoot over into each other as quickly and to remain in as unstable an
equilibrium, in the higher birds and mammals as in man.

WILLIAM JAMES

2.1 RANGE AND TERRITORY

An animal *taxon* (species, variety or other taxonomic group) has a
geographical range within which it occupies one or more habitats:
habitats occupied by rats include agricultural land and built-up areas.
A habitat may be divided into biotopes, such as hedgerows and warehouses. Within these are found colonies of rats; and each individual rat
in a colony has, over a given period, a certain range of movement. The
whole district over which an animal moves is its *home range*. If an
animal attacks or by other means drives away other members of its own
species in a part or the whole of its home range, the region in which it
does so is its *territory*. The attacks need not be, and usually are not, on
all other members of its species; but they may also be on members of
other species. The home range of wild rats is probably as a rule quite
small. It was studied by Davis and his colleagues in Baltimore and on a
farm: in the city they trapped, marked and released a number of rats,
and recaptured them; the distance between first and second captures
was less than 20 m for eighty per cent of the rats; they also put a dye
in rat bait, and found that the distribution of coloured dung around the
bait stations indicated a range of about 30 m in diameter [84]. Rats

are territorial animals, but for males the relationship of territory with home range is not known. For females the territory, when there is one, is the nest.

The main subject of this chapter is the movement of rats about their home range, together with the analysis of the causes of movement by means of laboratory experiment. A more fully documented review has been published elsewhere [21]. It is convenient to begin the detailed account of behaviour with this subject, since movement is involved in all behaviour; further, the study of movement in general illustrates many of the most important principles of ethology.

2.2 APPETITIVE AND EXPLORATORY BEHAVIOUR

2.2.1 *A problem of motivation*

Rats move about when they are getting food or water or nest material; when they are finding a mate or a site for nesting; and when they are fleeing from a predator. But their movements are often by no means obviously related to such activities. We therefore face at once a difficult question: to what extent are rats' movements determined by immediate need? And this obliges us to ask further: just what stimuli, external and internal, provoke and direct the movements?

It has long been known that laboratory rats are highly exploratory and 'inquisitive'. They are more obviously so than many other familiar mammalian species, because they have to move about to get information about their surroundings; others can do much by just turning their heads. One of the earliest studies of rat behaviour, that of Small, published in 1899, describes this behaviour at length and refers to the restlessness of infant rats as 'premonitions of curiosity' [291]. Wild rats, too, are actively exploratory, although this is often masked by a form of 'wariness' which laboratory rats do not display (§ 2.3). Generalized movement about a substantial area can, however, easily be seen in wild rats when an entirely strange place becomes accessible to them. This can be arranged by removing a barrier which has hitherto prevented the passage of rats living on ground adjacent to the new area; or it can be done by putting rats in a large and unfamiliar cage or enclosure.

Wide-ranging movements occur, not only when rats are in a strange place but also in familiar surroundings; the movements are then performed at a lower intensity, but they occur quite regularly. They may be observed even in a small cage. In examining the causes of these movements we begin with the internal ones. We saw in the previous chapter that activity is sometimes set going by an internal change, that is, a change of '*motivation*'. Is this the case with the exploratory movements of rats? Consider, for instance, rats which have been deprived of food for twenty-four hours. While they are asleep or resting in the back of their cages, food is put in the place where it is usually found at the front. Soon the rats become active and almost at once they begin to eat. When they have eaten a good deal and have drunk some water they carry out a general exploration of the cage [18]. Evidently, in this situation, satisfaction of a need (the attainment of the consummatory state of a full belly) *releases* exploration.

Of course, it could be that there was some special internal state, other than hunger, related to this outburst of exploration; and in any case it might be that in quite another sort of situation exploratory behaviour is a reflexion of hunger or some similar state. These problems have been partly solved by a great number of ingenious laboratory experiments on tame rats, some of which must now be described.

2.2.2 *The causes of activity*

It has long been known that *activity* increases in certain physiological states, for instance during oestrus. This can be shown in small cages, each of which communicates with a wheel or treadmill in which the rat can run. The revolutions of the wheel are recorded by a cyclometer. Rats run in the wheel very readily (figure 3).

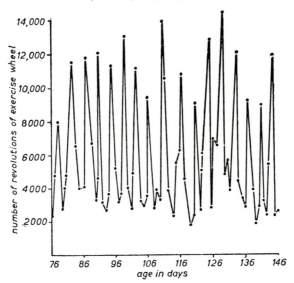

3. A behavioural cycle. The four-day oestrous cycle of the laboratory rat is accompanied by changes in the amount of running done in an exercise wheel. Peaks of activity accompany oestrus. (After Richter [255].)

While the early work on activity, notably that of Richter, emphasized internal stimulation, more recent experiments have suggested that external influences are also important. It has been suggested that activity depends on a changing environment, such as a laboratory in which there is some noise and movement always to be heard and seen. The increased activity in certain bodily states might then be due to a raised sensitivity to external stimulation. Hall tested rats in activity wheels (i) while in silence and in the dark and (ii) while subjected to external stimulation from flashing lights and noises; some of the rats were hungry, some sated. In these conditions rats did show increased activity resulting from external stimulation, but a still greater increase attributable to hunger.

Running in a wheel, however, though it certainly comes under the heading of activity, is not exploratory behaviour, and might not be related to exploration at all. In fact, although the use of running wheels has been quite a prominent feature of research on rat behaviour since the 1920's, there is still much that needs to be understood about the nature of the response to a wheel. The presence of a wheel, or at least its movement once running has begun, may act as a stimulus to further running.

Experiments with rats in wheels, or even in mazes to which they are transferred by the experimenter for each trial, are indeed very 'artificial'; that is, they depart in many respects from any of the kinds of situation in which wild rats are likely to have been at any stage in their evolution. However, other kinds of activity also have been studied. For instance, Fehrer observed rats in cages so arranged that, at a chosen time, the experimenter could give a rat in its nest box the opportunity to enter another compartment. When this was done with laboratory rats they showed a greater readiness to leave the familiar part of the cage if they were hungry.

In another investigation the movements of rats were observed in a cage with two compartments connected by three doors. To cross from one compartment to another, a rat had to find the one door which was unlocked. Two special circumstances were found to increase the number of crossings. One was the presence of a bright light shining on the rat; the other was hunger – or, more precisely, food deprivation. But even in the absence of either, many crossings were made. In these experiments, then, the energising effect of hunger was superimposed on a tendency to explore regardless of specific need. This conclusion accords with the observations on rats in activity wheels described above. However, in some conditions the exploratory behaviour of laboratory rats, like that of wild rats, is reduced by a state of need such as hunger (figures 4, 5).

In all this discussion one possibility has so far been ignored, namely, that rats tend to be active regardless of specific stimulation. This hypothesis might be expressed by saying that there is a 'general activity drive'. If so, it would be expected that, as with behaviour such as eating,

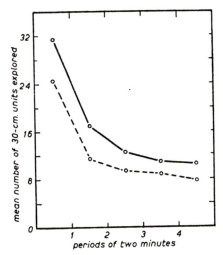

4. Effects of hunger and thirst on exploratory behaviour. Sated rats (continuous line) explored more than rats which had been deprived of food or water for some hours (broken line). In different conditions contrasted results may be obtained, as in the figure below. (After Montgomery [222].)

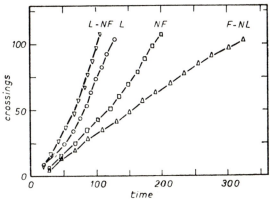

5. Effects of hunger and strong light on activity. Light (L) and hunger (NF) both increased the rate at which rats in one box crossed into another, connected box; but even the rats which had been fed and were not exposed to light (F – NL) crossed often. The graphs show mean number of crossings against time in minutes. (After Jerome et al. [163].)

deprivation would increase activity. Montgomery found that moderate confinement in a small cage did not affect the intensity of subsequent exploration in a simple maze [223]; but Hill later reported that confinement was followed by increased motor output when this was measured in cages which tilted with every movement made by the rat [144]. In monkeys (*Macacca*) Butler finds that a period of deprivation increases the tendency towards visual exploration [61].

On this question, then, we could do with more information. But the main conclusions from the researches described above remain unchanged. Activity in rats, measured in a wheel, in a maze or in a more open space, is increased by three kinds of stimulation. First, *changing* environmental stimuli, acting on the exteroceptors, increase activity. Second, some kinds of *constant* environmental stimulation, such as a bright light, do the same, since in general bright light is avoided. Third, deprivation, for instance of food, (*a*) lowers the threshold of response to external changes, (*b*) may itself increase activity in a constant environment.

2.2.3 *Spontaneous alternation*

There is however much more to be said on the external causes of exploration. These have been displayed especially in experiments on spontaneous alternation. This is most easily seen in mazes in which the animal is *not* rewarded on reaching a goal. The type of maze most used has only one point at which a choice of routes must be made (figure 6).

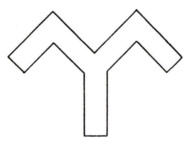

6. Plan of **Y** – maze. Each arm is about 30 cm long.

This makes it easy to express the animals' responses quantitatively. In such a T– or Y–maze, if rats are given two successive opportunities to make the choice, on the second occasion they tend to make the choice not made on the first run: if they turned left on the first run, they will probably turn right on the second. The same sort of behaviour can be observed in more complex mazes. Further, the complexity of the maze influences the intensity of exploration (figure 7). Spontaneous alternation may be regarded as a special case of exploratory behaviour, since it tends to increase an animal's range of movements. It follows, therefore, that an analysis of alternation is likely to throw light on exploratory behaviour in general.

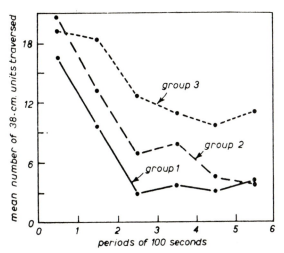

7. Effect of complexity on exploratory behaviour. Rats were put in an I–maze (group 1), a L–maze (group 2) or a T–maze (group 3): the more complex the maze, the greater the amount of movement. (After Montgomery [221a].)

An early hypothesis was that rats tend to vary the *response* made in a given situation: on this view, the performance of a movement induces a state of 'reactive inhibition' which for a short period prevents the repetition of that movement. The inhibitory effect, whatever its nature, dissipates gradually. For instance, in an investigation by Heathers one group of rats had an interval between choices of 15 seconds, while another group had to wait 120 seconds: in the first group alternation between the two possible paths occurred in 83·5 per cent of trials, while in the second the percentage was only 65·6.

Such behaviour need not, however, be due to an inhibition of the response: an alternative hypothesis is that the animal is varying the *stimuli* that it receives from one choice to the next. On this view the animal behaves as if it were bored with one part of the maze and therefore turns to another: it avoids not a recently performed response, but stimuli recently experienced.

Is this behaviour in fact response alternation or stimulus alternation? This has been investigated, by Walker and others, in +–mazes which can be rotated so that any of the four arms can point towards any feature of the laboratory; the arms of the maze are painted black or white, and any arm can be blocked, so that the + can be converted into a T. With this equipment the effects of three factors can be distinguished: (i) the response, that is, turning left or right; (ii) the visual stimulus provided by the differently painted arms of the maze; (iii) the visual stimulus provided by the background outside the maze. Thus on any given occasion the maze can be arranged so that a rat has a choice between alternating (i) its response, (ii) the stimulus provided by the background or (iii) that provided by the maze. In these conditions the alternation observed varied the visual stimulus received by the rat on successive choices, but there was no tendency to vary the response, that is, to alternate between turning right and turning left. The effects of the colour of the maze and of the laboratory background were about equal.

Nevertheless, other experiments have shown that in some conditions alternation may be determined partly by the actual movements made. A maze was used in which the entries to the goal arms were complicated: each involved twists of the rat's body and changes of direction, and the contortions required were quite different in the two arms. In these conditions the tendency to display response alternation was nearly as marked as that to display stimulus alternation. Evidently an important factor is the readiness with which two alternatives can be discriminated: merely turning left or right are not very different, but one set of contortions is easily distinguished from another.

An important question then is whether response alternation and stimulus alternation can both be considered as reflecting a single fundamental process. Dember put rats in a T–maze in which one goal arm was painted white and one black, but they were prevented by glass barriers from entering either arm on the first run. On a second exposure, both arms were now painted the same colour (*either* black *or* white), and either could be entered. Most of the rats explored the *changed* arm. This observation emphasizes the importance of *novelty*. It seems that alternation is an expression of a tendency to behave so that the input to the central nervous system is varied. This input may be through exteroceptors, as in the usual situation, in which the visual stimuli differ; or it may be through proprioceptors, as in the case of response alternation. This 'novelty principle' probably applies to all kinds of stimulation whatever. For example, rats given a choice of three or four foods nearly always sample them all during a feeding period, even if one is more readily eaten than the rest. The sampling tends to occur after a period of feeding on the preferred food.

2.2.4 *Attention to change*

Most recent studies, not only of spontaneous alternation but of exploratory behaviour generally, strengthen the conclusion that variety of the stimulus situation is important. If a laboratory rat is faced with a small enclosure in which there are various objects of differing degrees of familiarity, it will spend more time exploring new things than things already experienced. This decline in the 'interest' or 'attention' aroused is observed even when an object is presented only for the second time. If this response of rats can properly be described as one of boredom, these observations suggest that rats are very easily bored.

It is, however, more profitable to analyse further the precise features of the situation to which the rat is responding. When a rat enters, or is put into, an enclosure containing various objects, it approaches, sniffs and perhaps licks or gnaws them; it also moves around the whole enclosure. What exactly makes it do so? We have seen that *novelty* in the stimuli presented is evidently one factor. The importance of novelty alone has been most clearly shown by presenting laboratory rats with a variety of visual patterns. This was done by Montgomery by exposing cards with different black and white patterns on the walls of a simple maze. Given this stimulation, rats explore (that is, sniff and move around) more than control rats exposed only to a constant stimulus. However, another aspect of some of the stimuli that evoke exploration might be their *complexity*; and a third factor might be the *spaciousness* of the area available for exploration. A beginning has been made, by Berlyne & Slater, in the attempt to distinguish these three. First, rats in a Y-maze were found to sniff at an unfamiliarly patterned card more than at a familiar one; this confirmed what was already known. Next, one goal arm of the Y-maze was given a more spacious goal box at the end, and it was found that this was entered more frequently than a simple blind alley. Third, complex stimuli were compared to simple ones: this was done with cards, some of which were plain, others patterned; changing a pattern was more effective than merely altering from black to white or white to black.

Although novelty, complexity and spaciousness, separated in this way, seem very distinct, it is likely that they all act in the same way. A rat in a maze may be regarded as scanning the total situation, not only with its eyes but also with its other exteroceptors; the greater the complexity of the stimuli it scans, and the more spacious the area accessible to its senses, the greater the variety of input through its sense organs.

It might be thought that spaciousness at least was in a special category, since the tendency to choose the larger of two alternative areas might reflect an attempt to *escape*. This suggestion is worth discussing, since it illustrates a point of method. For a man, to escape is to move permanently from some confinement to a preferred place. But when a rat explores an area, familiar or not, near its nest, it typically returns in the end to its starting point. This is well seen in wild rats that have 'escaped', either by the design of the experimenter or against his wishes, from a cage or enclosure: they usually return to the accustomed nesting site, or other cover nearby, to sleep. If an animal permanently leaves its home ground, and one wishes to explain this, it is not adequate to say that the animal 'wished to escape': it is necessary to say *from what* it was escaping. This may be a state of deprivation, for instance of a mate; or it may be some positively disadvantageous feature of its home, such as continual disturbance by man or by a hostile companion. The suggestion of escape, unqualified, is in fact an anthropomorphism. It is based on the fact that most human beings tend to avoid being confined in a small area. By contrast, as Hediger [139] has pointed out, many animals, if provided with all their obvious needs, show no impulse to escape from cages or enclosures.

2.2.5 Exploration as 'reward'

The facts outlined above have led some psychologists to use expressions like 'stimulus hunger' and 'curiosity drive' to refer to the internal springs of exploratory behaviour. These terms may be compared with the expression 'investigatory reflex', used by Pavlov and his followers. There are many so-called simple reflexes, such as salivation, the eye blink and tendon stretch reflexes, which are evoked in almost any animal by an easily defined stimulus without previous learning by practice. These responses occur only when the specific stimulus is applied. If it is not applied for a long period there is no effect on behaviour: the animal does not become restless and move about until it can perform the response.

By contrast, as we have seen, rats do tend actively to seek variety in the stimuli falling on them. This has been confirmed in a number of experiments which show more clearly the difference between exploratory behaviour and automatisms such as simple reflexes. In these experiments, rats are placed in simple learning situations. A maze may be used. The task to be learned may be merely to turn left on each run in a T-maze, or to turn always towards some object, such as a light, regardless of whether it is on the left or the right. An animal is usually rewarded with food when it correctly performs the task to be learned. To ensure that the reward is accepted, the rat is deprived of food beforehand. In the experiments on 'stimulus hunger' the reward consists simply of an opportunity to explore. Turning in one direction leads to a plain blind alley, turning in the other leads to a larger area. In this situation the rats, instead of alternating, as they do when there is no reward in either of the alternative goals, learn to choose the path leading to the larger space. However, as with the experiments already described, spaciousness itself is not a *necessary* feature of the reward: merely presenting novel visual stimuli is also effective.

We shall see in § 7.3.2.1 that positive reinforcement may be of two kinds, primary or secondary. Food is an example of the first; opportunity to visit a place where food has previously been available could be a secondary reinforcement. Montgomery & Zimbardo have raised the question whether exploratory behaviour reflects the effect of secondary reinforcement, since in early life such behaviour must have preceded the finding of food; they suggest that as a result exploratory behaviour may be regarded as having been learnt at the time of weaning [224]. On this view exploratory behaviour should decline in intensity when it is not rewarded by food or something equivalent, that is, it should undergo extinction (§§ 6.2.1, 7.3.3); but it does not do so. This is a subject which needs further study: it is certainly still possible that secondary reinforcement plays a part in some circumstances. Meanwhile, the facts revealed in the experiments described above establish unequivocally that exploration has reward value of some sort.

Novel visual stimuli, and access to a large space, are not the only stimulus situations which possess this rewarding property. This has been shown in experiments in which 'Skinner boxes' were used (§ 6.2.3). The rat has access to a lever which, when pressed, releases a reward such as a pellet of food (figure 8): it soon learns to press the lever and to enjoy the fruits of doing so. Kish & Antonitis found that if the lever produces nothing more interesting than a clicking noise, this has a reinforcing effect: that is, it induces rats frequently to press the lever [174]. A quite different example has been provided by Kagan & Birkun, on whose rats the opportunity to take exercise in a running wheel was reinforcing [165]. It seems that rats tend to behave so that, within a given period such as twenty-four hours, they achieve a certain level of central nervous excitation via the exteroceptors. One way of testing this notion is to put one sensory modality out of action and to observe the result. Accordingly, Glickman blinded hooded rats, and found that they explored more than sighted controls [114].

In § 1 it was stated that the behaviour of an animal is often directed towards the achievement of a consummatory state, such as that produced by eating. Most of the obvious consummatory states are related to objects or conditions needed for survival, such as food or shelter. The work on exploration as reinforcement suggests that, in certain respects at least, variation of the stimuli acting on a rat may resemble a consummatory state. That is, rats are impelled to explore, in the same way as they are impelled to find and eat food. This is so much the case that in

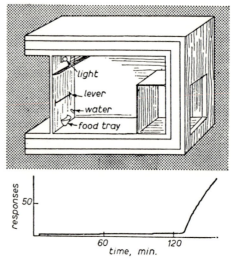

8. A Skinner box and a record obtained with it. The graph is a cumulative record of the number of occasions on which the bar was pressed, when every bar-pressing was rewarded. (From Skinner [290].)

some circumstances exploration may have priority over other behaviour that is usually strongly motivated. Wild rats ordinarily take cover when disturbed, but if they are placed in a completely unfamiliar enclosure they usually explore it before settling in a concealed place. This is not because they have had no previous experience of where cover is to be found: in large cages fitted with nest boxes, the boxes may be visited during the exploration; but the rat is so stimulated by the new surroundings that it emerges again, despite the presence of the experimenter, and resumes its wanderings. Later, when the conditions are familiar, exploratory movements are repeated at intervals, but only when there is no disturbance. Similarly, a male rat, placed in a cage already occupied by other males, may persist in leaving cover and exploring, even though it provokes repeated attack (§ 4.3).

Comparable observations have been made in more formal experiments on laboratory rats. Figure 9 shows a checkerboard maze, in each of whose squares there is a cup which may hold food or water. Rats were made thirsty or hungry, and then placed in the maze with water or food in the cups. The thirsty rats explored for many minutes before drinking,

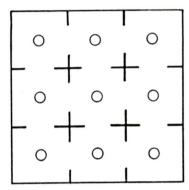

9. Checkerboard maze, with a cup in each unit, used for studying the effect of the presence of food or water on exploratory behaviour. (From Zimbardo & Montgomery [343].)

and hungry rats for still longer before eating. Another method of demonstrating the same principle is to use a maze with a conventional starting point and a goal, but so designed that the arrangement of the passage ways can be altered between runs. Figure 10 is an example: in

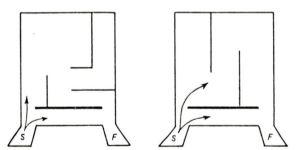

10. Maze which provides two alternative routes to the goal. If the longer one is varied (compare the two diagrams), a rat tends to prefer it to the shorter. (After Hebb & Mahut [137].)

this maze hungry rats tended to choose the *longer* of two alternative routes to food, provided that this route was *changed* between successive trials.

The character of exploratory behaviour is, then, amply demonstrated for rats: it is a response to variation in the surroundings, whether that variation is in time or in space and whether there is opportunity to move around in new areas or only to perceive new stimuli; further, there is a strong tendency actively to seek variety. Pavlov's term, 'investigatory reflex' is not very appropriate for such complex behaviour. Among animals more 'advanced' than rats exploratory behaviour has been studied in some detail in rhesus monkeys (*Macacca mulatta*) and, in less detail, in chimpanzees (*Pan troglodytes*). In man (*Homo sapiens*), it is more conspicuous than in any other species. Exploratory behaviour is also important in many, probably all, other mammals, though only in a few, such as the domestic dog (*Canis familiaris*), apart from those mentioned, has it been formally studied. There is a dearth of information on other vertebrates, though there is no doubt that birds are exploratory. For information on them, and on invertebrates, Thorpe [312] should be consulted. Apart from the researches mentioned by Thorpe, Darchen has shown that a cockroach, *Blattella germanica*, displays behaviour analogous in several ways to that of the exploratory behaviour of rats [79, 80]. Despite the small number of species studied, it seems that exploratory behaviour is a general phenomenon in the animal kingdom.

2.3 'NEOPHOBIA'

Wild rats are known to farmers, warehousekeepers and householders as wary and elusive animals which can be trapped or poisoned only with difficulty. It is not immediately obvious how this picture, based on millennia of experience, can be reconciled with that of the preceding section, in which rats appear as insatiably inquisitive and indeed as compulsively poking their noses into anything new. Certainly, if wild rats behaved in this way on all occasions, there would soon be few left in habitats occupied by man. In fact, they are protected by a type of response which is the opposite of exploratory behaviour and which sometimes completely inhibits it. This behaviour was first systematically studied in rats during field researches on methods of poison baiting. Its main features are seen in wild rats settled in an area in which they have established pathways. Use of such a pathway can be completely prevented, sometimes for days, merely by placing on it some unfamiliar object, such as a box (figure 11). Even a heap of nutritious food may have this effect for a time. This 'new object reaction' may be evoked even by a familiar object in a new position. Unfamiliar noises have a similar effect, but the extent to which wild rats become habituated to them has not been studied in detail. There is virtually no evidence that unfamiliar odours cause avoidance. However, there has been no rigorous work analysing the whole range of stimuli which induce avoidance. The question, What is, for a rat, a 'new object'?, has yet to be fully answered.

Accounts of the avoidance of strange objects will not be found in any of the published works on 'rat psychology': rat psychology is based on the study of strains of tame rats, and these do not avoid new objects.

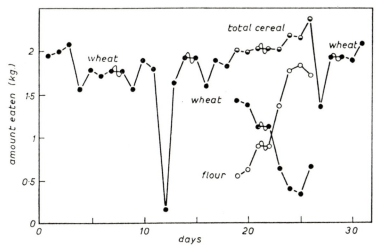

II. Avoidance behaviour of wild rats. The amount of food eaten by a colony of wild rats was recorded daily. From the twelfth day the food was placed on tin trays instead of the floor: this led to a sharp decline in food consumption during the following 24-hour period. Making flour available, as well as wheat, had no effect on the total eaten; but removal of the flour after eight days, when the rats had become accustomed to eating it, caused a decline for one day. (From Chitty [67].)

This has been clearly demonstrated by experiments in small cages, in which wild rats and tame rats (all born and reared in the same laboratory) were subjected to exactly the same conditions. If a wild rat is accustomed to feed from a wire basket at the back of its cage, and food is transferred to an unfamiliar tin in the front, the result may be a complete refusal of food for several days (figure 12). By contrast, tame white or hooded rats, faced with the new object, begin to explore it within a few minutes, and their daily intake of food is unaffected. The feeding of a wild rat may even be interrupted if an unfamiliar tin is placed in the front of the cage, without any change in the position of the food; and this interruption is not due to the rat being distracted by exploring the new object: on the contrary, the tin is for some time completely avoided. In these experiments, then, food consumption is used as a convenient index of 'neophobia'; and in the second kind of experiment it seems that the strange object by itself induces a state of 'fear' which by itself inhibits feeding even on the familiar food in its familiar place.

Two important features of avoidance behaviour or 'new object reaction' must be emphasized. One is that it is temporary: although the effect can last for weeks, usually the avoidance is overcome in a few days or even hours. This is an example of habituation (§ 6.2.2). For habituation to take place, the new situation must of course remain unchanged: any other alteration in the environment will drive the rats away again. These facts have obvious implications if one wishes to deal with an infestation of wild rats, for example by trapping.

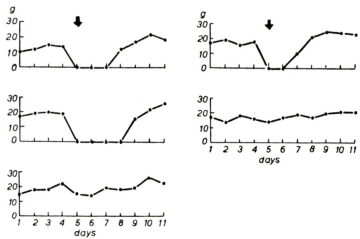

12. 'Neophobia' in wild rats. Daily consumption of food of five adult *Rattus norvegicus* each alone in a small cage. After four days food was supplied in an unfamiliar container; three of the rats then stopped eating altogether for two or more days and the feeding of all was interrupted. This avoidance of novelty is not shown by laboratory rats. (From Barnett [20].)

The second general feature of 'new object reaction' is of greater importance for understanding its nature: it occurs always when there is a *change in an otherwise familiar situation*. In wild rats, as we have seen, investigation dominates in a totally new environment, while avoidance (flight, 'fear') results from a relatively minor change in a familiar constellation. The important factor seems to be contrast between what is expected (that is, what has been experienced before) and what is actually observed. Hebb gives an example of neophobia, similar to that of rats, in his study of fear and anger in chimpanzees. Of the stimuli to which these responses are made he writes: 'An essential feature of the stimulating conditions is the divergence of the object avoided from a familiar group of objects while still having enough of their properties to fall within the same class' [131]. It is possible that this sort of behaviour is widespread at least among the mammals, but much remains to be learnt about the form it takes in different species. Moreover, it is not universal: in the house mouse (*Mus musculus*) it is replaced by a capricious and unpredictable kind of behaviour; and it is evidently absent from the behaviour of shrews (*Sorex*).

2.4 THE ECOLOGY OF EXPLORATION AND AVOIDANCE

2.4.1 *The functions of exploration*

The discussion so far has been mainly on the stimuli which evoke approach or avoidance. We must now turn to the survival value of these kinds of behaviour.

Wild rats, once they leave the nest, are subject to predation, and their principal means of avoiding predators are the use of pathways under cover and flight to a burrow or other place of concealment. This behaviour depends on previous experience of the topography of their living space. Given this, they can run from any one point to any other, by the shortest route and in the least possible time. Rats are more vulnerable to predation by cats if they are in unfamiliar surroundings. But exploration does not lead only to learning where cover and nest sites are: it is accompanied by sampling of all the edible or potable materials encountered, and so a rat is regularly informed of sources of food and water. A further consequence of general movement is that odour trails are left on the routes regularly used, and these are detected by other rats (§ 4.2.1.4).

The word 'random' is sometimes applied to exploratory and appetitive behaviour. A more appropriate term would be 'unpredictable', but even this is misleading. The direction of a rat's movements, even in an unknown terrain where learning cannot have occurred, is influenced in an orderly way by many of the features it encounters. The inorganic stimuli may be tactile, visual or olfactory. (The vibrissae, as Vincent [325] has shown, play an important part in receiving tactile stimuli.) A similar variety of stimuli may be provided by other rats.

Rats, like many other small mammals, tend to move in contact with a vertical surface; they also eat in a corner rather than an open space. This behaviour is an example of thigmotaxis, or movement orientated by tactile stimuli; it is one of the few examples of specific orientations encountered in rat behaviour: its exceptional character illustrates one feature in which mammalian behaviour differs from that of most invertebrates. Thigmotaxis, formally studied in young rats by Crozier

[77a], has usually been regarded as behaviour which appears regardless of the conditions in which the rat is reared or of any particular sort of experience. On this view it would be called 'innate' or 'not learned'. But Patrick & Laughlin have reported that rats raised in an environment without opaque walls do not develop a tendency to move close to a wall [240]. This is an example of difficulty in determining how a behaviour pattern develops and to what extent it is independent of individual experience (§ 5.3.2). Functionally, thigmotactic behaviour is presumably an aspect of taking cover from predators. This is more obviously the case with the tendency, already mentioned, to move from light to dark. Other guiding stimuli are more likely to lead to exposure: the odours of food or of other rats have an attractive effect which may overcome the shelter-seeking tendency.

The exploratory movements of rats in any ordinary environment have, then, superimposed on them, effects of external stimulation which make the direction of movement much more predictable than it would be if it were influenced only by the novelty of the surrounding objects.

2.4.2 *Avoidance behaviour*

We have seen that, if wild rats in human communities behaved as tame ones do towards unfamiliar objects, their compulsive exploratory and sampling behaviour would quickly lead them into traps or result in the ingestion of poison bait. 'New object reaction', displayed in a highly developed form by wild rats, protects them from the consequences of curiosity. It seems likely that the extent to which rats commensal with man display avoidance behaviour is a consequence of selection over the seven thousand years during which civilization, with its stores of food, has existed. More intense selection *against* wildness and neophobia produced the laboratory rat in perhaps half a century.

However, avoidance of strange objects, and especially strange animals of the same and other species, is common in the animal kingdom. In many species of birds and mammals it develops early in life, after a brief period when the young become imprinted (§ 8.4.1) on their parents. Once this young-parent attachment has been acquired the safety of the young is well served by their avoidance of other animals. The avoidance behaviour of wild rats has perhaps a source in this sort of behaviour.

Whatever its evolutionary origin, the neophobia of wild rats is not by itself sufficient protection against poisoned food. It is combined with a capacity to learn to refuse toxic mixtures. This parallels the ability, much studied in tame rats, to select, in some instances, the nutritionally superior of two foods (§ 3.3.1). The combination of exploring and avoidance with learning is therefore elegantly adapted to giving a rat a maximum of information about the resources and dangers of its environment, in the safest possible way.

2.5 CONCLUSIONS

Many animals regularly explore their surroundings: that is, they approach and enter every accessible place in an area or volume around a nest or resting place. Exploratory behaviour is one category of the variable movements of animals. Sometimes, such movements end when a specific (consummatory) state is achieved, whether by performance of a consummatory act or not; or, if movement does not cease, a new kind of activity follows. It is often possible to assert, or to infer, that the activity has been evoked by a specific internal state, such as hunger. The term 'appetitive', applied to this sort of behaviour, is most conveniently used to distinguish variable (unpredictable) movements from stereotyped activities such as courtship ceremonies; the latter are not variable but fixed (predictable in detail) within each species.

The definition of 'appetitive behaviour' thus becomes 'any variable behaviour'. Whenever it is studied it is desirable to state to what 'goal object' (if any) and to what internal state (if known) it is related. Some exploratory behaviour is unrelated to any goal and, probably, independent of any special internal state.

Exploratory behaviour in satiated rats is a clear example of variable movements unrelated to immediate need. It is seen most readily in laboratory rats, but is highly developed also in wild rats. It is not a spontaneous motor activity, but a result of a general tendency to vary the stimulus situation. Other species, from cockroaches to man, behave in a similar way.

Wild rats given access to an unfamiliar area explore it. They also regularly re-explore their home range. The value of this behaviour is that it provides information about the resources and dangers of the environment. But movement in a familiar area is inhibited by the presence of an unfamiliar object or arrangement. This neophobia is not found in laboratory varieties. Its function is to protect rats from danger. Wild rats thus possess a delicately balanced and highly efficient combination of movements of approach and avoidance, which enables them safely to make the most of what the environment has to offer.

3

Feeding Behaviour

Not one man in a billion, when taking his dinner, ever thinks of utility. He eats
because the food tastes good and makes him want more. If you ask him *why* he
should want to eat more of what tastes like that, instead of revering you as a
philosopher he will probably laugh at you for a fool.

WILLIAM JAMES

3.1 GENERAL

Much of the behaviour described in the preceding chapter is only
indirectly related to the primary needs, that is, needs which must be
satisfied if the animal is to survive for more than a few minutes, hours
or days. The internal processes which determine that animals will
behave so as to satisfy these needs are sometimes called the 'homeostatic drives'. As we saw in § 1, such processes are set in motion by
some internal disturbance: for parameters such as the concentration of
sodium ion in the body fluids, the amount of carbon dioxide in the
blood or the temperature of the skin, there is a narrow range of values
which the animal tends to maintain. This chapter deals with the
behaviour which maintains the rat's intake of food and water.

3.2 THE COMPONENTS OF FEEDING BEHAVIOUR

3.2.1 *Directional movements*

3.2.1.1 *Olfactory stimuli.* It was shown in the preceding chapter that
the variable movements of rats tend to put them regularly in every
accessible spot in a substantial area around their nest. This enables
them to learn where food and water are, to strengthen that learning at
intervals, and to take advantage of the appearance of new sources of
food and water. These variable movements are supported by the
sampling of all materials encountered. The stimuli involved are largely
olfactory and gustatory. In the rat, as in most mammals, the olfactory
sense plays an important part in behaviour. The structure of the brain
reflects this olfactory dominance (figure 13). It seems likely, though it
is not proved, that odour leads to the first sampling: any of a wide range

of odours seems to induce licking, and this is usually followed by actual eating.

The effect of odours in attracting laboratory rats has been studied by Neuhaus. He points out that a rat is continually sniffing and turning its head; there must consequently be rapid variation in the intensity and character of the stimuli influencing the olfactory organs (yet a further

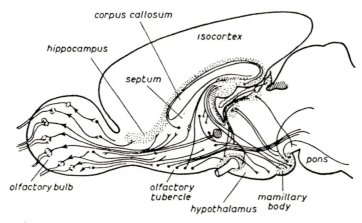

13. Olfactory and related connexions in the rat brain. (After Herrick [142].)

aspect, perhaps, of the tendency to vary the sensory input as much as possible). Neuhaus found that the direction of the movements of rats is influenced by air currents which carry odours. He put rats where they had a choice of two air currents, bearing different odours, and trained them to go to food marked by the odour of formic acid; after this training, the rats continued to do so when given a choice between the odours of formic acid and another, such as acetic or valerianic. On the other hand, rats trained to approach formic acid, and given a choice between this and butyric acid, went to the latter, and had to learn *not* to do so. Thus in these experiments butyric acid acted as a 'lure' [234]. (For wild rats, by contrast, butyric acid tends to act as a deterrent [32].)

Eayrs & Moulton have made a more thorough study of olfactory acuity in laboratory rats. They find that in determining the direction of movement, odours are subordinate to other sensory cues and that the olfactory stimulus should be close to the reward (for instance, food) if learning to respond to the stimulus is to be efficient [89, 229].

These experiments illustrate three points. First, odours can determine the direction of a rat's movements. Second, there are probably fixed responses (approach or avoidance) to some odours – responses that may well be general to a species or variety regardless of any special experience; this, however, needs to be tested. Third, rats can learn to associate food with particular odours, and even with an odour to which they had previously shown an aversion.

3.2.1.2 *Social stimuli.* The direction of a wild rat's movements as it goes to eat may also be influenced by other rats. In this book any such effect is called 'social'. No detailed experimental work has been done on social influences on feeding in wild rat colonies, but some observations have been made on artificial, enclosed colonies.

Probably most, if not all, social interactions in feeding behaviour are due to learning. For example, Barnett & Spencer watched colonies of wild rats when they began to feed in the evening. The food was wheat grains. Usually, one particular member of the colony emerged first, and carried a mouthful of grains back to its nest; the nest was shared with other rats, and these came out soon after the return of the 'pioneer'. The possible importance of such rats appeared in two kinds of situation. In one, the rats were deprived of food for one or two days, and then food was replaced in the usual containers. The first rat to find that food was available again was the pioneer of the colony, and the return of this rat to the nest with food led to a rapid and unanimous sortie by the rest of the colony. This occurred, as one would expect, especially when the rats had had previous experience of a period of starvation followed by a return of food. In a second kind of experiment, the food put in the enclosure was of a kind that was only rarely available to the rats. This was done from time to time with cabbage. The first rat out at once took a cabbage leaf and returned with it to the nest. This was followed by a great deal of activity, including excursions for additional leaves and attempts to wrest fragments of leaf from other rats.

The habit of taking food to the nest, and eating it under cover, had other indirect social effects. Grains dropped from a rat's mouth, as a preliminary to eating them one by one, were often taken by other rats. If a rat had been eating flour, residues on its face or hands were licked off by other rats, especially young ones [18]. Chitty [67] has seen young rats turned on their backs and robbed by older rats, on returning to cover with a mouth full of grains. In no instance did theft lead to conflict. In every case, the probability is that rats have learned to associate certain behaviour of other rats, or certain appearances or odours, with the presence of food. The fact that a social interaction is involved is incidental.

It is often suggested, nevertheless, that rats are coöperative animals, and band together to carry food to their nests. There is a persistent legend that eggs are taken in the hands of one rat which then lies on its back while a second rat drags it by the tail to cover. This belief should not be casually dismissed: rats do use their hands for grasping, and they do pull each others' tails. It is not impossible that rats could learn, almost accidentally, to perform some such trick. However, in what seems to be the only authentic account of egg-stealing by rats with supporting photographs (plate 1b), the rats operated individually, each pushing an egg along in front of it. The observations of Barnett & Spencer already quoted suggest that even *learning* to coöperate in food-getting is exceptional. When rats found a lump of liver, weighing 450 g, outside the nest, the larger ones dragged it under cover (plate 1a); sometimes two or three were seen heaving away at one time. But they did not pull in the same direction: on the contrary, they often pulled in opposite directions [31].

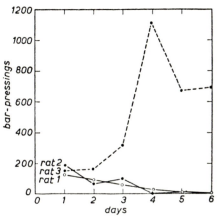

14. 'Coöperative' behaviour. Three laboratory rats were put in a cage (a Skinner box) in which pressing a bar at one end released food at the other; there was no other source of food. For two days little bar-pressing was done but, after this, one rat developed the habit of working for all three. (After Mowrer [231].)

It is possible in rather special conditions to induce laboratory rats to adopt a form of 'coöperation'. Mowrer has described experiments in which three rats were put together in a modified Skinner box (compare figure 8, page 26); pressing a lever at one end of the cage released food at the other end. At first no rat did much lever-pressing, evidently because doing so led to no reward for the performer but only an unearned pellet of food for one of the others. Eventually, however, *one* of the rats developed the habit of doing enough lever-pressing to satisfy the needs of all three (figure 14). The apparent parallel to one kind of human community was in this case a result of a comparatively simple learning process. An interesting question, left unanswered by Mowrer and his colleagues, is what determines that one rat should become a worker and the others mere 'parasites'.

In some mammals, notably Carnivora, Proboscidea and Primates, the young receive maternal guidance. There is no evidence of this in rats. Young rats often follow their mother, at least while they are still feeding from her; but a female with active young does not, except incidentally, guide them to food. She may indeed push them out of the way if they obstruct her while she is eating [18]. As mentioned in § 2, as soon as their eyes open young rats begin independent exploratory forays; before they are weaned they consequently have had experience of a substantial area around the nest. It is, evidently, in this way that they learn their way about, and not by any social process; they do not learn by imitating their mother as kittens (*Felis*) do. However, a detailed experimental study remains to be made.

The conclusion on social effects on feeding, then, is that they reflect rats' ability to *learn* to respond to all kinds of stimuli, including those provided by other rats; but that no more specific social interactions are involved, except of course those of a mother with sucking young.

3.2.2 *Avoidance and its aftermath*

The preceding paragraphs are concerned with the approach to food. It has already been described how wild rats, although they eventually sample everything within range, at first avoid an unfamiliar food, or familiar food in a new place (§ 2.3). There are two aspects of the effect of 'new object reaction' on food consumption. First, an environmental change may cause an interruption of feeding even though the food is not changed or moved (figure 11, page 29). This evidently represents a generalized state of 'fear', as described in the previous chapter. Second, a change in the conditions at a feeding point may reduce the amount eaten at that point, without necessarily influencing eating elsewhere.

We have seen that neophobia wanes, as a rule, quite rapidly. If a new food becomes available, increasing amounts of it are eaten, until it may even be preferred to what was eaten before (figure 19, page 52). When the amount of food is less than the rats can eat, a further effect is observed. Thompson has studied this in colonies living in natural conditions; some of the rats had been caught, marked and released, so that they could be individually observed. Figure 15 shows the progressive concentration of feeding near the time when food was made available each evening by an experimenter; in these conditions, the intensity of feeding is increased, and its duration reduced. Here the behaviour is adapted, not only to a topographical situation, but also to a time relationship. We shall see below that the behaviour of rats towards new foods protects them very efficiently from poison.

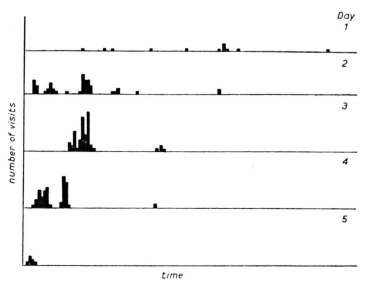

15. Decline of avoidance of unfamiliar food with time. Each histogram records the number of visits of a single marked wild rat, living in a colony on a pig farm, to a pile of food. The record for each day begins at 17.00 hr. when the food was put down. Food had not been available at this place before the first day of the experiment. On the fifth day poison was added to the food and this rat, together with most of the others, was killed. (From Thompson [309].)

3.2.3 *Eating*

Among the components of feeding behaviour we have still to deal with the act of eating itself. It would be absurd to leave this out, even though there is remarkably little to be said about it.

Eating is the 'consummatory act' of feeding behaviour. It involves licking, chewing, salivation, swallowing and handling. The first of these is an important part of sampling. Chewing, salivation and swallowing are usually considered to be 'reflex' activities, and are therefore sometimes disregarded by students of behaviour. Handling may be discussed more fully. It is a feature of the behaviour of all rats. Manipulatory behaviour has been seen in wild rat nestlings as early as thirteen days, before the eyes had opened: they took fragments of wood in their hands and, in a clumsy manner, conveyed them to the mouth [26]. Barnett has also watched newly weaned wild rats, feeding for the first time on flour or sugar: they begin by burying their noses in the food but, as a rule, within a few minutes, they use one or both hands, at first intermittently but soon consistently. For some weeks, nevertheless, young rats, like babies, continue to spill a good deal of food, and they only · gradually come to feed with little waste [18].

Among adults, although handling always occurs, there is much individual variation in the form it takes. Typically, a rat rests on its hind quarters and uses both hands (figure 16). Spencer has pointed out

16. Adult, male *Rattus norvegicus* eating food held in the hands.

that the stereotyped nature of this behaviour even extends to the precise way in which a cereal grain is nibbled: while smaller rodents, such as mice, *Mus musculus*, hold a grain at right angles to the long axis of the body, and eat it in the way in which a man eats corn on the cob, a rat holds it as a man holds a cigar [299]. This, however, may be due to learning which is the more convenient way, mechanically speaking, of eating grain, and the difference between mice and rats may be merely a function of size. Certainly, if one watches wild rats eating flour, one sees no uniformity in the detail of the movements: one may squat and use two hands in the manner described; another may put its nose well down into the food; a third may sit up and use one hand only, as a scoop.

Although, then, handling itself is species-characteristic and *seems* to be fixed or 'innate', its details are variable. We cannot at present identify the factors involved in its development. No doubt there are features of the structure of the limbs, and in the workings of the central nervous system, which are present in all rats and which enable them to use their hands; but the extent to which individual experience plays a part, and the exact nature of the relevant stimuli, have still to be discovered. This kind of question is further discussed in § 5.

3.2.4 Carrying and hoarding

We have seen that, when rats have made their way to food, they may take the food under cover before eating it (plate 1). Carrying food under cover may happen even when the rats have been deprived of food for some time. Wheat grains are taken away in mouthfuls: even a rat of medium weight (250 g) will take six to eight grains at a journey; a large rat takes more, and may be seen shovelling them rapidly into his mouth until the cheek pouches are full [31]. Under cover, the grains are disgorged and eaten one at a time. This behaviour no doubt helps to protect rats from predators.

Sometimes food is taken to a nest or burrow and deposited without being eaten. (Other objects, such as stones, cakes of soap, bits of wood and much else may be accumulated in the same way [26]. We have no explanation of this behaviour, and it has been largely ignored in experimental studies of hoarding.) In stable colonies of wild rats kept in large cages (§ 4), food cubes are sometimes removed as soon as they are put in the food boxes, and scattered on the floor or taken to the nest boxes [19]. Calhoun, from a study of wild rats in a large enclosure, concluded that most rats store food in their burrows, but especially those which have been mildly attacked by other rats. He reports also that rats which had been severely attacked took food to scattered points nearby and then paid no further attention to it. He suggests that nevertheless this behaviour has a biological function, since scattering food makes it more easily accessible to rats living at some distance from the main source [64]. This hypothesis is not proved but, if it is true, scattering the food would favour the growth of a colony by reducing conflict.

Hoarding occurs also in tame rats, and in them it has been the subject of many experiments, of which the early ones have been well reviewed by Munn [233]. It is a common observation that tame rats, given objects such as food cubes in an open container, at once pick them up and scatter them around the cage. In a cage with a covered refuge they tend to accumulate objects in this 'home' box. Morgan and his colleagues found that this happened especially if the rats had recently been deprived of food for a time [226]. This is not a direct result of hunger: the hoarding does not begin as soon as the rat is hungry, and it goes on long after the rat has been able to eat its fill – sometimes even for twelve weeks; and the pellets are typically not eaten, but merely deposited in the living cage [225]. The *biological* significance of this behaviour, as we have just seen, may easily be surmised; but even if we knew for certain that hoarding had survival value, either for the individual or for the colony, the *immediate stimuli* which influence it would still be open to investigation.

The fact that pellets are collected in the nest suggests that it is their presence which constitutes a consummatory state: that is, that the goal of the behaviour is the hoard itself. If this is so, removal of the pellets as they are brought back should increase the number brought in while, presumably, deliberately stocking the home cage with them ought to inhibit the behaviour. The result of removing pellets as they are brought in varies [211, 212]. If the rats are not hungry, hoarding occurs whether the cubes are removed or not; but if they are hungry, removal of the cubes during preliminary training in the cage leads to an appreciable *reduction* in the number of cubes hoarded in a given time when this is subsequently tested. Thus the inhibitory effect of removal of pellets depends on the extent to which the rat needs food at the time when it first experiences the hoarding situation. 'Disappointment' at this stage leads to subsequent decline in the impulse to carry out an act which has proved fruitless on previous occasions.

Since rats are liable to take any portable objects to their nests, it may be asked whether they would hoard just as much if the pellets were not edible. Miller & Vieck have shown that, just as previous food deprivation may increase food hoarding, so previous water deprivation may increase the hoarding of water. This was shown by supplying pledgets soaked in water, instead of, or in addition to, food pellets. When given a choice, food-deprived rats chose pellets, water-deprived rats took pledgets. Small wood blocks, offered as a further control, were taken little or not at all.

The same authors have investigated the characteristics of a home cage that induce rats to hoard there rather than anywhere else. Familiarity was found to be an important factor, especially the familiarity of its odour [215, 324]. Another factor is the amount of cover it provides: Bindra found that accumulation of pellets in the living cage is increased if the alley leading to the food tin is open, instead of being closed at the top. Bindra observed greater hesitation in entering the open alleys, and interprets his results in terms of the greater 'security' offered by the home cage [47]. This notion requires further analysis. However, the importance of a home in which the rat is out of sight of enemies, has contact stimuli from the walls, is protected from draughts and bright lights and is surrounded by familiar conditions and odours is very evident.

The facts given so far are enough to show that hoarding is a complex behaviour pattern analogous to the accumulation of food stores by squirrels (Sciuridae) and beavers (Castoridae). The universality of the behaviour in rats has led to the use of the term 'the hoarding instinct' [225]. If this expression is intended only as a name for a commonly observed type of behaviour there is no great objection to it; but if it is intended to convey that the behaviour is 'innate', or that it is universal in rats, or that it develops independently of a rat's individual experience, then it is misleading. This is well shown by the analysis of hoarding behaviour made by Marx [210], on which the following account is largely based. He points out that rats used in the experiments described above have usually had considerable experience of food pellets, and so may well have been influenced by this experience. Further, an understanding of the way in which hoarding develops in an individual rat requires one to consider the several, partly separate, elements which make up the whole pattern. First, there is exploration; second, there is homing or taking cover emphasized in Bindra's work quoted above; third, there is the series of actions evoked by the pellet; seizing it (a preliminary to eating); carrying it (a result of taking cover); and releasing it (before taking it in the hands to eat). The total pattern of hoarding is regarded as a result of the fusion of these elements, each of which is held to be an independently acquired habit. The primary reward which leads to learning of the habits (apart from taking cover) is the opportunity to eat, since any seizing or manipulation of a food object early in life is usually followed by eating. Soon the smell and taste of food pellets come to act as secondary rewards or reinforcements, as a result of association with the primary satisfaction of eating.

This is at least an attempt at a detailed analysis of a complex stereotyped action pattern, and it shows how involved the interaction between nature and nurture must be in the development of such an activity (§ 5).

3.3 FACTORS INFLUENCING CHOICE OF FOOD

3.3.1 *Physiological effects of food*

3.3.1.1 *Favourable effects*. We turn next to the question of choice of foods. Two relevant facts have already been mentioned: first, that rats which have experienced thirst hoard water rather than food, and food-deprived rats food rather than water; second, that if rats, wild or tame, have access to two or more foods, they do not ordinarily restrict themselves only to one, but at least sample all of them. These facts suggest that the exploratory behaviour and learning capacity of rats combine to enable them to make the best use of whatever is available around them. We shall see that, with some qualification, this is indeed true.

That rats are exceedingly versatile in their food habits is one of the most familiar facts about them; it has obviously contributed to their success both as pests and as laboratory animals. It also makes them convenient animals for the study of food selection. The fact that they are liable to sample everything makes it necessary to express their preferences in terms of the proportion of different foods making up the whole diet: among a wide range of foods choice is a quantitative affair. This raises the question, in what units should the quantities be expressed? Sometimes a convenient method is to give the calorigenic value of each food. This is because rats, like other animals, regulate their eating to maintain a fairly constant intake in terms of utilizable energy. The last statement requires one major qualification: food consumption is a function of body weight: a rat grows throughout life and so the amount of food it eats rises with age. This effect is small in the adult, since growth is slow and the increase in food eaten per unit of body weight declines with increasing weight. Another way of expressing this relationship is in terms of calories per unit surface area. Harte and his colleagues showed that, expressed in this way, the intake of laboratory rats rises during the second week after weaning (at three weeks); after this it decreases, until it levels off at about 13 kcal/dm²/day. In the adult, regardless of body weight, energy intake is a function of surface area [128].

The constancy of energy intake is the most obvious example of the influence of the physiological effects of food on food selection: the amount of a given food eaten depends, other things equal, on the energy value of that food and of the other foods eaten during the same period. Hausmann added ethyl alcohol to the drinking water of rats and found that the intake of other sources of calories declined in proportion to the calorigenic value of the alcohol consumed. He also compared the effects of adding sugar and saccharin to the drinking water: the saccharin had (for man) the same sweetening effect as the sugar but no energy value; and intake of solids declined proportionally when sugar was added but was unchanged with saccharin [129]. Similarly Adolph added undigestible 'roughage' to rats' diet, and the volume of food they ingested increased so that its energy value remained constant; only when the proportion of nutrients in the mixture was reduced to one-third or less was the intake too little to maintain body weight. Adolph also found that when food and water were given together, in the form of milk, the volume drunk was determined by energy needs alone, despite the excess of liquid that had to be taken [2]. Energy intake remains remarkably steady despite variations in external temperature and the composition of the diet. Sellers and his colleagues kept rats at $1.5°C$ and varied the protein and fat content over a wide range without upsetting calory consumption [282, 283].

Calorie intake is thus independent of the growth-giving quality of food, for example its protein content: growth is, from this point of view, an incidental consequence of consuming food that gives a specific amount of energy. Aschkenasy-Lelu [14], and Scott & Quint [275], have studied the effects of protein deficiency in albino rats: given a choice between a diet short of protein and one rich in protein, rats do not adjust their intake to their need; some eat enough protein, others do not, and it seems impossible to predict which will happen.

Although protein has no special effect there are individual substances which can influence food choice as a result of their nutritional value. The behaviour involved is sometimes called 'dietary self-selection'. The earliest important study was by L. J. Harris and his colleagues in 1933. They were primarily concerned with the physiology of nutrition, but they were faced with the fact that what an animal (or a man) eats is not only a matter of food chemistry, but depends also on factors which may be unrelated to any obvious biochemical need; they were therefore led into observations of behaviour and into asking what was meant when an animal's choice of diet was said to be 'due to instinct'. They made rats deficient in vitamin B_1 (thiamin), and gave them a choice between two diets of which one was flavoured with Bovril, the other with Marmite; the latter, a yeast extract, contains B vitamins. Control rats (not deficient) ate the diets without discrimination, switching readily from one to the other, unless and until they developed a deficiency; the deficient rats consistently ate the mixture containing B vitamins. If a choice of several foods was offered they sometimes failed to identify the one with the favourable effect; but rats which had had previous experience of the 'correct' mixture ate it even when presented with a large choice. Rats familiar with the choice situation 'give a cursory sniff at each food in turn and then proceed to eat the vitamin-containing diet'. The effect of thiamin on the heart rate was detectable an hour after its ingestion, and Harris and his colleagues attributed the correct choice of diet to association between this favourable physiological action and eating a particular mixture. They considered that labelling the diets with a distinctive flavour was important [126].

These remarkable observations have been confirmed and extended in a number of studies: among them are those of Aschkenasy-Lelu [15], Richter [255], Scott, and Young [339]. Scott and his colleagues made albino rats deficient in thiamin, riboflavin or pyridoxin; such rats chose a mixture containing the vitamin they needed, in preference to a deficient diet, even when the mixtures were not labelled with flavours [274]. Thiamin-deficient rats were not infallible in these conditions, but those that did succeed readily picked out the one vitamin-containing diet out of four offered [277]. Tribe & Gordon, in carefully designed and analysed experiments, have even been able to show that rats which are not thiamin-deficient have a slight tendency to prefer a thiamin-containing diet [319]. But not all vitamin deficiences are similarly met by altered behaviour: Young & Wittenborn for instance could not detect any 'self-selection' effect in rats needing vitamins A or D [341]. This may be related to a slower physiological action by these substances [272].

The results with B vitamins are paralleled by some on the selection of inorganic salts, of which sodium chloride has been most studied. Richter and his colleagues have done many experiments which showed increased salt consumption by salt-deficient rats, and especially by animals from which the adrenal glands had been removed. The absence of adrenal cortical tissue causes a profound disturbance of salt metabolism: much sodium chloride is excreted in the urine, and a corresponding increase in dietary salt is needed to maintain life. Adrenalectomized rats, given a choice of plain or salty water, consume much more of the latter than do control animals; they may consequently survive in good health [255]. However, Scott and his colleagues found that rats made salt-deficient merely by being fed on an almost salt-free diet showed no corresponding change of behaviour towards solutions containing salt. Such rats do not need as much salt as those that have no adrenals, and it may be that they took in enough salt to counteract the deficiency without showing a significant *preference* for the salt solution: that is, they drank without distinction both the salt solution and the plain water offered, and in this way received the salt they needed incidentally [278]. The physiology of the regulation of sodium chloride intake is discussed further in § 3.4.2 below.

Another example, due to Richter, of self-regulation of intake of an inorganic salt is found in the behaviour of rats after parathyroidectomy. This operation greatly increases the need for calcium, and accordinglo the rats show an increased appetite for calcium lactate [255]. This was confirmed by Scott and his colleagues [279].

There is scattered evidence, reviewed by Lepkovsky [195], on the selection of advantageous diets by other mammals, including man. It seems likely that the phenomenon is a general one. It depends largely on the rapid learning of the favourable effects of particular foods. Animals other than mammals no doubt adjust their intake to their need for energy, but nothing seems to be known of more refined forms of dietary selection. It may be that this is an example of the adaptability of behaviour which is a special feature of mammals.

3.3.1.2 *Unfavourable effects.* The selection of one food in preference to another involves at least the partial rejection of the second food. Consequently, it may not always be clear whether it is the favourable effect of the food selected, or the unfavourable action of the food refused, that is significant. Figure 19, page 52 gives an example which also illustrates the possible added complexity of preferred flavour (see § 3.3.2 below). The workers whose researches were described in the preceding paragraphs usually assumed that the more nourishing food was not only biochemically better but also had an internal effect which gave it 'reward-value', in the sense in which that term is used in studies of learned behaviour: that is, ingestion of the better food produced a con-summatory state. This attitude is supported by the fact that the less nourishing foods are accepted if no better alternative is offered: there is no evidence of an aversion to these foods. However, Barnett & Spencer have shown that rats made hungry will accept flavoured food which, when offered with an unflavoured alternative, is completely rejected [32]; it follows that failure to reject a food is not complete evidence for its general acceptability.

There are, as already mentioned (§ 3.2.2), clear instances of learned rejections. These come from study of the effects of poison baits used in attempts to kill wild rats in places such as farms and warehouses. We have seen that wild rats avoid new things, including foods, in a familiar environment. Chitty and his colleagues showed that, after the stage of total avoidance of the bait, there is a tendency to sample it: as fear wanes, small amounts of the food are taken. It is as if fear and curiosity are in conflict, the first gradually declining and giving way to the second; yet a single act of sampling seems to result in a temporary decline in curio-sity (or an increase of fear), since it is typically followed by a period in which the food is not eaten [67, 267, 268]. The anthropomorphic terms in which this behaviour is here described could be replaced by expressions such as 'approach', 'withdrawal' and so forth, but the actual behaviour would remain the same. We do, however, need a much more detailed analysis than we have at present, not only of the factors which cause the fear or withdrawal, but also of the way in which these factors interact with those which provoke exploration and sampling. Such a study would have to be carried out on wild rats (plate 2).

The interval after the first sampling of a new food gives an opportunity for the substances it contains to act on the body. If there is a poison present, it may begin to take effect and so put an end to feeding altogether for a time. If the rat recovers, as often happens when only a small amount of poisoned food has been taken, it will then probably refuse the poisoned food on subsequent occasions [267]. This is evidently the obverse of the dietary self-selection described above: the rats have learnt to reject instead of accept.

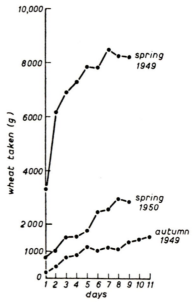

17. A method of census. Each graph shows the daily consumption of food by rats in a small English village. The amount eaten rises steadily to a plateau which provides an index of population size. After the first census, in spring 1949, a thorough poisoning campaign was carried out. The other two graphs represent the rat population level six months and one year later respectively. (After Barnett, Bathard & Spencer [28].)

This sequence of responses is extraordinarily effective in protecting wild rats from poisoning. Chitty has described the results of a large series of experiments in the field, designed to find out accurately how many rats are killed by putting down poisoned food at a large number of points in an infested area. The rat populations were in each instance estimated by a technique of indirect 'census'; this involved putting down (usually not at the poison points) weighed amounts of wheat grains, and recording the daily consumption until a plateau was reached, as in figure 17; this gave a measure of the rat population present. After the poisoning, another census was carried out. Of 58 experiments, in which zinc phosphide, Zn_3P_2, or alphanaphthylthiourea were the poisons, only just over one-third gave a reduction of 85 per cent or over of the rat population; these figures illustrate the protective effect of avoiding new objects, sampling and learning to reject injurious foods [67].

The poisons used by Chitty and his colleagues were not among the most toxic available. Zinc phosphide, for instance, has a LD50 of about 41 mg per kilogram body weight: that is, administration of this dosage to a large number of rats may be expected to kill half of them. It might be supposed that much more toxic poisons would kill much larger percentages. However, Barnett & Spencer used sodium fluoroacetate, which has a LD50 of about 3·8 mg per kilogram for laboratory rats and is possibly even more toxic to wild rats. Even with this poison, there were failures. One consequence of unsuccessful poisoning is, as already mentioned, that the rats, after recovery, tend to refuse the food that has caused illness. This 'bait-shyness' was found in colonies which had been subjected to sodium fluoroacetate, as well as less drastic poisons, showing that the rats had sampled and so learned to avoid the highly toxic bait [30].

Sodium fluoroacetate acts quickly, and this was no doubt a factor in the development of poison-shyness. Accordingly, it might be supposed that a poison with a slow action would take effect insidiously, and would not cause shyness. However, as already mentioned, arsenious oxide, As_2O_3, which may act quite slowly, undoubtedly causes shyness. Further, Armour & Barnett were able to bring evidence of shyness to a poison which acts cumulatively, as a result of its repeated ingestion over a period of several days. This poison, dicoumarin (3·3'-methylene-bis-4-hydroxycoumarin) gradually reduces the clotting power of the blood, and eventually causes internal bleeding. Baiting large infestations of wild rats with food mixed with dicoumarin led to a reduction in the numbers of rats, but the effect fell short of complete destruction: after some weeks, an equilibrium was reached between the lethal effects of the baiting, and breeding by the rat population (figure 18). This was evidently due to the development of bait-shyness among the remaining rats.

Experiments in the field throw light on the ways in which behaviour contributes to survival, and provide at least indirect evidence on the responses of rats to foods; but they do not give conclusive evidence for phenomena such as bait-shyness. For instance, poisoning causes a *general* refusal of food soon after it has taken effect; this is not bait-shyness, since the latter involves discriminating a food that has caused illness from other foods. General avoidance behaviour is indeed, as Rzoska has shown, one feature of the behaviour of rats that have eaten poisoned food. Another difficulty is that in the field it is usually

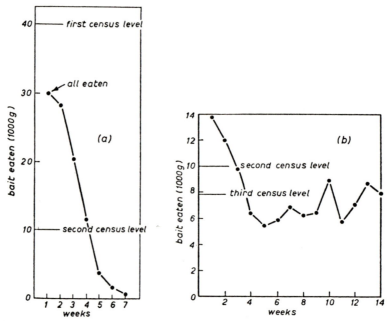

18. Effect of a cumulative poison. Graph (*a*) shows the weekly consumption, by a large colony of wild rats, of a bait containing dicoumarin, a blood anticoagulant which acts as a cumulative poison. Before the poison was first put down a census indicated a population of at least 1600 rats; the poisoning evidently reduced this number by about 75 per cent. Graph (*b*) shows what happened when the residual population was given the same poison in a different bait base: there was little effect in 14 weeks, evidently because the rats had developed 'poison shyness'. (After Armour & Barnett [12].)

impracticable to study rats individually; and the survivors after a poisoning may be exceptional rats not to be compared with those that have succumbed. However, the authors quoted above have also carried out laboratory experiments, and these confirm that rats do learn to refuse foods that have made them ill. Further, Rzoska has shown that the rats' learned aversion may be displayed to only one component of a mixture which has previously had a toxic effect: if a bait containing sugar has poison added to it, and rats survive the poisoning, they may subsequently refuse other mixtures containing sugar. These facts apply both to wild rats and to albinos [267, 268].

The central facts, then, on food selection are (i) the ability to modify behaviour in accordance with the physiological effects of food; and (ii) the tendency to avoid the unfamiliar in a known setting regardless of specific experience. These, together with exploration and sampling, enable a rat to make considerable adjustments of its feeding behaviour according to circumstances.

3.3.2 *Palatibility and aversion*

It must not be thought that a rat's choice of foods is uniformly determined by its homeostatic needs, and that when a rat fails to perform like a dietician this is highly exceptional. Rats often make their choice on an entirely different basis: their behaviour resembles that of a man when we say that he likes or dislikes some taste. The qualities of foods discussed in § 3.3.1 influence the amount of a food eaten by means of a change in the animal's behaviour: the change, whether it is an increase or a decrease in the amount of food eaten, is a consequence of a physiological effect of the food which can be objectively assessed as favourable or unfavourable. But some physiological actions of foods do not lead to a change in the response to the food itself: they induce either acceptance or refusal from the first, and continue to do so. These effects, evidently of flavour or texture, *seem* to be due to an organization of the chemical senses and the central nervous system which is independent of the differences of environment to which individuals are subjected. This, however, is only an hypothesis which needs to be tested by experiment.

An example is that rats often prefer soft or finely divided foods to harder or coarser ones. Carlson & Hoelzel found that laboratory rats tend to eat the softer part of grains and to leave the harder; but if grains are soaked in water they are eaten whole [66]. Soaking, however, may influence taste, through the formation of sugars. The effect of texture is most clearly shown where the only difference between two foods is their state of division. Colonies of wild rats, observed by Barnett & Spencer, preferred wholemeal to whole wheat grains; since the wholemeal had been made from the wheat, the state of division was evidently responsible for the choice. When one of these colonies, previously accustomed to eating whole wheat, was offered wheatmeal as an alternative, there was a gradual switch, lasting one to three days (figure 19), before a steady intake was again attained.

A probably similar effect is brought about by the presence of oils or fats in food mixtures. This has been studied for albino rats by Scott &

Verney [276], and for wild rats by Barnett & Spencer [33]. Both kinds of rat eat mixtures of carbohydrate foods with oils of not very marked taste (for instance, arachis oil) in preference to the plain foods. Scott & Verney consider that this is an effect of palatability, and not of nutritional value. Some other oils are refused, or eaten in only small amounts. Cod liver oil is accepted only in small quantities by wild rats, but it is just possible that this is due to a toxic effect of the vitamin A it contains. Corn oil, cottonseed oil and butterfat are not eaten well by albino rats.

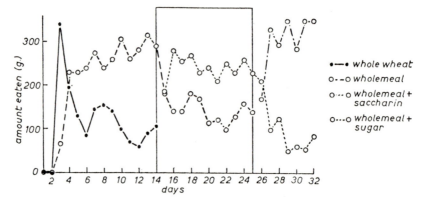

19. Effects of state of division and of sweetness on food choice by a colony of wild rats. Food was entirely refused for the first two days, because the food was offered in an unfamiliar container. On the third day whole wheat, which was familiar, was preferred; after that there was a change to meal, which is ordinarily preferred to whole grains by wild rats. Later observations showed the effects of adding (*a*) saccharin and (*b*) sugar to the meal. (After Barnett & Spencer [33].)

The effect of sweet flavour has been analysed in more detail than those of texture or of oils, and it illustrates the difficulties of interpreting preferences that at first sight seem simple. The facts we have concerning wild rats may be given first. In the colonies, already mentioned, set up by Barnett & Spencer, wheatmeal mixed with sugar was preferred to wheatmeal alone. Sugar of course has a rapid effect on an animal's internal state. However, saccharin had a similar effect (figure 19). These observations suggest that foods containing sugar or saccharin are preferred simply because of their taste, that is, for their direct action on the gustatory organs. Nevertheless, this conclusion, on the facts, could be regarded only as an hypothesis, for the following reasons. If rats have been accustomed to sweet foods (of which the maternal milk might be one), a sweet taste would possibly be *associated* with food and its favourable internal effects, and so come to be the cue for a learned preference. In this case the sweet taste would be a secondary reward. However, Sheffield and his colleagues carried out experiments on white rats in which the animals were required to discriminate between the taste of saccharin solution and that of plain water. The sweet taste acted as an effective reward in a learning situation: the rats learned to go to one side of a T-maze when doing so enabled them to drink the sweetened water. These authors concluded that the taste of saccharin was *not* an acquired reward, for two reasons. First, their rats had had no previous experience of anything so sweet and so, it was held, could not have learned to associate sweetness with a primary reward. Second, the effect of saccharin persisted, however long the rats were tested; whereas, if the saccharin were effective only through its association with some other effect, such as that of food in the stomach, it would be expected to lose this effect with repeated presentations: that is, extinction of the response should take place [286, 287]. This conclusion has not gone unchallenged. Smith & Capretta have more recently brought evidence that saccharin solutions act as rewards only when they have been associated with the presence of sugar in the stomach [293a]. This contradiction has yet to be resolved. The difficulties of studying apparently 'inborn' preferences and other characters are referred to again in § 5.3.

There are responses to flavours which seem much more obviously 'inborn' than that to sweetness: these are to substances which, mixed with acceptable food, cause complete refusal if any alternative is available. Wild rats display such aversions more readily than laboratory rats. Scott & Quint, for instance, did not deter albinos by adding aniseed oil or butyric acid to their food [273]; but Barnett & Spencer found that both of these repelled wild rats. Barnett & Spencer used boxes with perforated zinc lids to test responses to offensive odours: the food was put on the lid of the box, and the odorous substance inside. A powerful odour, that of n-butyl-mercaptan, acted as a repellant: most of the rats fed from a similar box nearby that contained no odorous substance. Other stinks, less offensive to man, of substances which act as deterrents when actually mixed with food, gave less decisive results [32].

Much remains to be learnt about the responses to tastes and odours. Among the problems is the extent to which apparently fixed responses can be modified by experience. Barnett gave young wild rats wheat mixed with cod liver oil or aniseed oil as their first solid food, with no alternative available. Later, he offered the familiar mixture (which the rats had been eating regularly), and wheat mixed with arachis oil as an alternative; the young rats almost at once transferred their attentions to the new, less noticeably flavoured grains [18]. These experiments show that there are aversions which can be nullified by hunger, but that there is no permanent alteration of behaviour: given the opportunity the rats turn to the usual choice. However, nothing has been done yet to test whether still earlier experience, in the period when milk is the only food, can influence the preferences shown in later life.

3.3.3 *Social facilitation*

The word 'social' is used here to refer to any behaviour which is directly influenced by another member of the same species (§ 4.1). In this sense, social behaviour may have an important influence on the *place* at which wild rats feed. The observations of Calhoun [64] on effects of conflict on feeding were mentioned above (§ 3.2.4), and the consequences of territorial conflict are dealt with further in the next chapter.

Here we are concerned with the circumstances in which rats stimulate other rats to feed, or to eat more than they would otherwise have done. Both of these may be brought under the heading of 'social facilitation' – a term defined by Crawford as 'any increment of activity resulting from the presence of another member of the same species' [76]. A special case of this is where rats seem to imitate other rats by feeding where the latter are already feeding. True imitation is rare in species other than man, and the apparent imitation is due to associative learning. It has been studied in laboratory rats by Miller. The rats were trained on a raised T-maze (figure 20) to turn towards either a white or a black card; the cards were on the right or the left, in a random sequence. A trained rat was put on the short arm of the apparatus, and behind it another rat; the second rat was rewarded with food, by raising the hinged lid of a sunken cup, if it followed, or 'imitated', the first rat. In these conditions the second rat might be learning to respond to the card; control trials were therefore made without cards, and with the leading rat either untrained or trained to turn in a specific direction. Figure 20 gives some results. These observations illustrate how easy it would be, in more 'natural' conditions, to mistake the results of ordinary associative learning for imitation. Since human beings imitate each other, we unthinkingly expect members of other species to do the same.

Another way in which social facilitation of feeding may come about has already been described in § 3.2.1.2, in the account of feeding in colonies of wild rats.

Although these examples of social facilitation may be accepted as results of fairly straightforward learning, there are others which are more obscure. Harlow found that laboratory rats in groups of two or more ate more than when they were alone. The experiments were carefully controlled, and suggested that the extra eating resulted from a 'competitive situation' [122]. Other, rather inconclusive work on feeding has been reviewed by Munn [233], together with some rather more convincing studies of the enhancement of drinking by social effects.

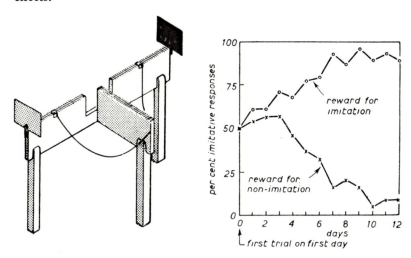

20. 'Imitation'. Elevated T–maze used for studying a form of learning which resembles imitation. Two rats were started on the short arm. Reward could be given by opening a sunken cup, containing food, in one of the transverse arms. Some rats were rewarded for following ('imitating') the first rat, others for not doing so. The graphs show the result. (After Miller & Dollard [219].)

The Soulairacs find that choice of food, as distinct from amount consumed, may be influenced by the presence of other rats. They confirm that laboratory rats in groups of two or three eat more than solitary individuals; but if they are offered both a standard solid diet and a glucose solution, the rats in groups eat more of the former but drink less of the solution than the solitary ones; however, the net result is still a greater intake of calories for the grouped rats. The average increment was as great among paired rats as among those in groups of three, but the increase among the former did not occur as soon as they were put together, but only after about twenty days; however, when a third rat was added to a pair, there was an immediate increase in consumption, followed later by a smaller decline [297]. Interpretation of these observations is not yet possible, and much more, and more detailed, work is needed before such social interactions can be understood.

Despite the obscure character of some social effects, none of them supports the notion that rats have any complex coöperative social behaviour. It has often been claimed, in anecdotes, that adults guide young rats to food, and even warn them of the danger of poison bait; but close watching of the behaviour of wild rats, in both natural and artificial colonies, gives no evidence of anything of the sort. We have seen that a female may be followed by her young, especially if they are still sucking, and so may incidentally guide them to food; but their exploratory behaviour will bring them to food without this help, and probably usually does so. Evidently, the most clearly defined social influences on feeding are the consequences of learning; and, although there are other, more obscure effects, in rats it is the learned associations that are biologically important.

3.4 INTERNAL MECHANISMS

3.4.1 *The problems*

Most of the preceding account of feeding behaviour has been in terms of the overt behaviour of rats faced with various situations, and of the environmental conditions, including the qualities of various sorts of food and the presence of other rats, which influence the behaviour. The internal processes involved have been merely mentioned, usually only by implication.

In the rest of this chapter we shall be concerned with those internal mechanisms which are special to eating and drinking. This means that we shall not be concerned, for example, with the neural changes on which learned behaviour depends, even though – as we have seen – much of feeding behaviour depends on learning. What little we know of the nervous system in learning is deferred to § 9. The specific mechanisms to be discussed here are of two kinds. First, there are 'peripheral' states, for instance in the stomach or in the composition of the blood. These may be regarded as the antithesis of the 'consummatory state': they help to make the animal active, and to remain active until they are changed in a certain way, in other words, until a consummatory state is reached. They constitute one aspect of the physiology of 'drive' or 'motivation' (§ 8). Second, parts of the nervous system are concerned in a very intimate way with eating and drinking: local damage to the brain may alter, say, feeding behaviour in a drastic way, without having any other evident effect. We then say that these parts 'control' eating and drinking, though it is not easy to give any more precise meaning to this assertion than the statement already made: that damage to them deranges the behaviour. The tendency to eat or drink, the intensity with which these activities are carried out, and the length of time they go on before stopping – all these are influenced by neural function; hence this too is an aspect of motivation.

3.4.2 *Thirst*

Adolph and his colleagues have reviewed the factors which influence the intake of water in rats and rabbits. They divide the whole process into four elements: the first is seeking water; the second is the act of drinking; third, cessation of drinking at a given moment; fourth, the absorption of water and its distribution to the tissues. Clearly, these could be further subdivided.

Seeking water has, in effect, been discussed in §§ 1.3 and 2.2, under the heading of appetitive movements. The internal processes which evoke appetitive behaviour, and thence drinking, are also responsible for the *act of drinking* itself. Adolph and his colleagues show that drinking occurs only if the total body water is below a certain level (that is, if there is a water deficit), or if there is an excess of solute in the body, such as may arise if sodium chloride, NaCl, is ingested. Either of these constitutes an internal stimulus to drink. Administration of NaCl was used to study in some detail the nature of the stimulus constituted by excess of salt. The salt influenced drinking only after it reached the blood. In the blood it brought about changes in at least three parameters: (i) the osmotic pressure of the body fluids; (ii) the concentration of chloride in the blood; (iii) the volume of the cells of the body (which tends to decrease with increasing tonicity of body fluids). By comparing the effect of urea with that of NaCl, evidence was brought that all these factors contribute to the impulse to drink. Cessation of drinking could be induced by distention of the stomach, as a result of its being either filled with fluid or being distended by means of an inflatable balloon on the end of an oesophageal tube. Although these authors assert that, 'within certain limits', the amount of water drunk is proportional to the magnitude of the deficit or the excess of solute, they also found that neither copious drinking nor a high rate of chloride excretion in the urine follows the production of a 'hyperchloric' state in a rat: evidently, something prevents a rat from *rapidly* compensating for an excess of solute in the body; but it is not known what this is [3].

Work by McCleary has, however, some bearing on this question. He gave laboratory rats glucose and fructose solutions of varying strengths, after a period of deprivation of water. The largest amounts of these solutions were drunk when they were at a concentration of 5·3 per cent, that is, isotonic with body fluids. The lower intake of higher concentrations is attributed to their effect in the stomach: they cause withdrawal of fluid into the stomach and so bring about dehydration elsewhere in the body. This suggestion was tested by introducing fluids of various tonicities directly into the stomach before the rats were given a solution to drink. The fluids put in the stomach were solutions of glucose, saccharin or NaCl, and it was found that their effects on subsequent drinking depended on their tonicity. By contrast, intraperitoneal administration of glucose solutions did not lead to a reduced ingestion of glucose solution. This was held to be due to the different effect on the blood: intraperitoneal glucose does not provoke hypertonicity of the blood, but there is a lowered blood plasma volume. McCleary suggests that there are two kinds of thirst: in one, glucose is acceptable; in the other, it is not. The difference depends on the state of affairs in the stomach [201].

It seems that, while the internal stimuli which provoke drinking are physico-chemical qualities of the body fluids, the cessation of drinking probably depends on receptors in the alimentary tract. McCleary gives evidence for the existence of such receptors in the stomach. More recently, Miller and his colleagues have brought evidence for their presence also in the mouth or pharynx. These workers gave water by stomach fistula to thirsty rats and found (as had previous investigators) that this had an immediately satiating effect: that is, it reduced the tendency of the rats to drink. However, they found that the same amount of water taken by the mouth was still more satiating [220].

The importance of signals from the alimentary tract is also illustrated by the work of Deutsch & Jones. They point out that rats given a choice between water and hypotonic saline drink more of the saline. This has been supposed to reflect a 'preference' for the taste. Yet they showed that rats offered hypotonic saline and water as alternative rewards in a T-maze learned to go to the *water*. Their explanation is based on the fact that the sensory nerve fibres which are influenced by water or saline in the mouth have a high rate of spontaneous firing; water reduces the rate of firing while saline has a smaller effect: from this point of view saline is diluted water. They suggest that a rat continues to drink until a certain amount of water has been signalled and that more saline has to be drunk than water to achieve this [85].

All the work described in the preceding paragraphs was done on adults. We do not know what makes an infant rat stop drinking milk. It is a reasonable hypothesis that the cessation of drinking in the adult is a result of learning: a certain quantity of stimulation in the mouth, pharynx and stomach may make a rat stop drinking, as a result of association of this level of sensory input with internal consequences of drinking. This kind of effect will be further discussed in the next sub-section.

3.4.3 'General hunger'

3.4.3.1 *Stimuli from the gut*. The knowledge we have of the control of eating suggests conclusions similar to those on drinking. The internal processes involved are multiple [117]. Again there are the divisions of food seeking, ingestion, cessation of eating and absorption. The internal causes of the first two (the appetitive and consummatory behaviour) are the same. The change from appetitive to consummatory occurs when the animal reaches food, that is, in response to an external stimulus.

What determines that a rat, previously quiescent, should become active, move towards food and eat it? Observation of rats in an environment which provides sleeping places and a reliable source of nutrition-ally adequate food shows that eating occurs at fairly regular intervals. The amount eaten, as we saw above (§ 3.3.1.1), enables the rat to main-tain a steady though diminishing rate of growth during most of its life. The cycle of feeding activity has been reviewed by Munn [233]. An early study was that of Richter, in 1927. White rats with access to food were found to have a cycle of three to four hours, and feeding coincided with the later part of a period of maximum general activity (cf. § 2.2.2). Powelson found that the activity cycle was paralleled by one in the muscles of the stomach: he used an operation in which the stomach of the rat was shifted to a position between the skin and abdominal wall; here its movements could be both seen and automatically recorded. Gastric peristalsis reached a peak at the time of greatest locomotor activity; sometimes the peristalsis began before the general movement, but more often the rat started to move around before the peristalsis began. The overt activity is in fact certainly not dependent on the gastric movements, at least in adults, since the activity cycle persists even if the stomach has been surgically removed, or the nerves to the stomach have been cut. Feeding too persists after these operations. This work does not, however, exclude the possibility that impulses from both the stomach and from other organs, particularly the intes-tines, play some part in inducing feeding behaviour: it shows only that the stomach does not play a *necessary* part.

There is indeed good evidence that impulses from the gut do influence eating, but its end rather than its beginning. Smith & Duffy have ex-amined the question whether the *bulk* of the stomach contents influences feeding. In their experiments rats were allowed to feed only during two hours out of each twenty-four, and they studied the effects of various previous treatments on the behaviour of the rats during the two hours. To some rats they administered a mixture of kaolin and water, to others, kaolin and glucose solution isotonic with body fluids. Both mixtures depressed eating, and there was no significant difference between their effects. We have seen above (§ 3.3.1.1) from the work of Adolph, that rats can adjust their intake to allow for excessive solid or liquid bulk, but that is a separate point. The work of Smith & Duffy suggests strongly that in the conditions of their experiments the rats were responding to increased volume of the stomach contents by ceasing to eat [293].

The influence of stimuli from the alimentary tract can also be illustrated by comparing the effects of taking a food through the mouth with those of having it injected into the stomach through a fistula. Kohn measured 'hunger' by the readiness of rats to operate a panel which, when pushed, released a drop of liquid food. He studied three groups of rats: one had 14 ml milk injected straight into the stomach through a fistula; a second had 14 ml saline injected in the same way; and the third were allowed to drink 14 ml milk in the ordinary manner. 'Hunger' was found to be lower after stomach milk than stomach saline, and lower still after mouth milk. Since the test for hunger was applied 5·5 min. after ingestion of the milk, the effect was rapid (figure 21).

3.4.3.2 *Internal states.* None of these experiments helps us to understand the *onset* of feeding. A common-sense hypothesis is that an animal eats when it needs food. This notion has two aspects. First, it might imply the more colloquial statement that an animal eats when it is hungry. A man knows he is hungry by his feelings, but we know nothing, and can surmise only a little, about the feelings of rats: if we wish to speak of a rat's being hungry we are obliged to base such a statement on observation of a rat's behaviour or on some 'physiological' criterion. The relevant behaviour consists of moving about, and of eating when food is reached. In an experimental situation, such as a Skinner box (figure 8, page 26) the behaviour could include pressing a bar or other learned act. Hence the word 'hunger', in ethological statements, could be defined either as the tendency to eat, or if preferred, as the internal state which tends to induce eating; in any case, the events referred to in the definition should be of a kind that can be observed by anybody. Whatever the exact definition, talk of hunger provides no explanation of a rat's behaviour: it does not help one to predict what a rat will eat or when, but merely *names* a type of behaviour or an internal state.

The second aspect of the notion, that a rat eats when it needs food, is heuristically more useful. It entails asking: what is meant by 'needing food' or, what is the criterion of 'need'? For example, if the concentration of glucose in the blood falls below a certain level, the animal becomes comatose. It is also true that the blood sugar level is fairly rapidly affected by eating. It is therefore reasonable to suggest that this might be the parameter which determines the onset of feeding.

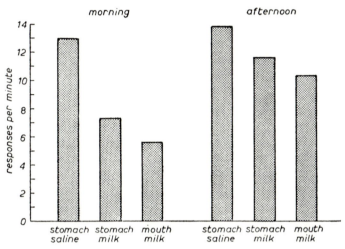

21. 'Hunger' and its satisfaction. Rats in a cage could obtain food by pushing a panel; this response was used as a measure of hunger. The different satiating effects of saline injected directly into the stomach, milk injected into the stomach and milk taken by the mouth are compared on two occasions in one day. (After Kohn [175].)

This is the basis of the glucostatic hypothesis of Mayer. It has been tested in various researches, of which a recent example is that of Smith & Duffy already quoted. They introduced a glucose solution, isotonic with body fluids, into the stomach by a gastric fistula; in this way the effects of taste and swallowing, if any, were excluded. No depression of eating followed [293]. Another way of getting glucose straight into the body is to inject solutions intraperitoneally. This has been done by Janowitz & Grossman; they found that it led to a slight depression of food intake, but not more than followed control injections of saline of the same volume. They went on administering glucose in amounts equivalent to 25 per cent of the energy needs of the rats for up to seven days, and still the animals failed to adjust their feeding to the excess of carbohydrate they received [161]. The blood sugar hypothesis, then, is contradicted by the experimental evidence, at least as far as rats are concerned.

An alternative proposal, made by Kennedy, is that the relevant internal state is the fat balance. Young rats maintain their stores of adipose tissue in a remarkably constant state, apparently as a result of sensitivity to the concentration of metabolites circulating in the blood. Kennedy's hypothesis is a 'lipostatic' one [170, 171].

A third suggestion, made by Brobeck, has been briefly expressed as follows: 'animals eat to keep warm, and stop eating to prevent hyperthermia' [56]. Brobeck has shown that the food intake of rats declines with rising environmental temperature, until above 34°C they develop a slight fever and almost completely stop eating, even though their metabolic rate is raised at this temperature. The suggestion is that they could not, in these conditions, stand the additional heat stress that would result from the specific dynamic action (SDA) of ingested food. As a test of the hypothesis the satiety value of different diets was related to their SDA. The hypothesis requires, for instance, that a diet containing much protein (with its high SDA) would be accepted in smaller amounts than a high fat diet. The results agreed with the hypothesis [58].

Among the questions which must be asked, when we consider the body temperature theory, is what significance should be assigned to the evidence, given above, for the influence of stimuli from the alimentary tract on feeding. The probable answer is that the effects of bulk in the stomach, and of tasting and swallowing, are the results of associative learning. We have seen that these can bring about a decline in eating before any foodstuffs actually enter the blood; but in the life history of an animal the gut stimuli must have been regularly followed by the primary actions of foodstuffs within the body. These actions include the effect on body temperature. It is possible that there is a neural organization, which develops regardless of individual experience, and which determines the different responses to rising or falling body temperature. A rise in temperature may come to be associated with the prior experience of food in the mouth and in the stomach; and so the stimuli from the gut come to evoke the response which previously followed a rise in temperature, namely, cessation of eating. On this assumption learning has in fact taken place in a quite straightforward way.

The account of Brobeck's proposal has allowed a discussion which at least illustrates the sort of physiological analysis that must be involved in accounting for the fact that an animal starts to eat and later stops.

There is, however, no convincing evidence that the releasing or un-conditional stimulus (§ 6.2.1) in this case is a change in body tempera-ture. Soulairac [296] may be consulted for the evidence against the hypothesis. The releasing stimulus might instead be gastric. We have seen that stimuli from the stomach are not needed for typical feeding behaviour in the adult; but they may be essential for the early learning of eating habits [248]. This is the suggestion made by Hebb [134] in a valuable discussion of hunger and learning.

3.4.3.3 *Defining 'hunger'*. It is now necessary to say something further about the meaning of the word 'hunger'. Hebb has made an important statement on this subject, though in a logically defective way. He begins by defining hunger as the tendency to eat – a definition, however brief, in terms of behaviour. Later he asks his readers to 'consider hunger to be . . . an organized neural activity' [133]. This is an example of the tendency to allow a term to slide unobtrusively from one meaning to another which has bedevilled so much writing about behaviour (and about other subjects too). The important contribution here is that Hebb directs attention to the neural activities which underlie feeding behav-iour; he points out that an organized neural activity can be aroused, or quelled, in more than one way as a result of learning: owing to associa-tion, stimuli from a variety of sources can come to activate a pattern of central nervous activity which leads to, or inhibits, feeding.

Another example of confusion on the meaning of the term hunger may be found in a valuable review by Miller of the relation of food consumption to other indices of feeding behaviour. His formulation is that food consumption is not a satisfactory measure of 'hunger', but he does not define 'hunger'. His true point, which is an important one, can be conveyed by quoting experiments to which he refers. In one kind of situation hungry rats are trained to press a bar which delivers a food pellet into a dish below. When a rat has been trained, the mechanism is set to deliver a pellet only irregularly and unpredictably. Miller writes: 'The animals continue working much like a gambler who operates a slot machine in the hope of hitting the jackpot. The rate at which they work seems to be a good measure of the strength of hunger.' The implication of the last sentence is that there is some process or state, which *could* be directly measured, but which is conveniently measured indirectly by the rate of bar-pressing; this state is named hunger.

Miller describes a second sort of situation in which rats are presented with very small samples of food, each one containing a slightly higher concentration of quinine than the last. To quote: '. . . the hungrier animals progress further up the series into the higher concentrations of quinine.' Miller then refers to the behaviour of rats which have lesions in a particular region of the hypothalamus (§ 3.4.5); these rats eat more food than normal ones, and become obese, but they press a bar for food *less* vigorously than normal rats and they are *more* easily deterred by the taste of quinine. Clearly, bar-pressing and the quinine technique measure one thing, food consumption measures another. Whether one says that one or the other experimental procedure measures hunger depends in the first place on what meaning one gives the word 'hunger'.

Miller describes further experiments which illustrate the difficulties of interpreting studies of the internal states that determine behaviour. Rats had two stomach fistulae, one opening into the stomach, the other into a balloon in the stomach. When milk was injected into the stomach, bar-pressing to get food was reduced; when fluid of equal volume and specific gravity was injected into the balloon (thus increasing the volume of the stomach contents but not the food supply) bar-pressing was again reduced, though not quite so much. Hence both might be said to have a satiating effect. Similar rats were trained in a T-maze: one group was rewarded, on turning to a given side, by injection of milk into the stomach; the other group received an injection of fluid into the stomach balloon. The first group duly learned to turn to the side where the milk was given; but the second group learned to *avoid* the side on which the balloon was inflated. In other words, the inflation of the balloon was punishing, not rewarding. Miller suggests that the reduction of bar-pressing which followed inflation of the balloon was not an effect of satiation but rather a consequence of nausea [218].

The general conclusion from these observations, and indeed from all those quoted in this and preceding sections, is not that food consumption is an inadequate measure of hunger, since that begs the question of what the term 'hunger' means: the conclusion is that there are many internal processes which influence feeding behaviour. There is no single process or state which can conveniently be named 'hunger' or 'hunger drive'; the internal springs of feeding behaviour are diverse, some acting together, others in opposition. This conclusion will be expanded at the end of the chapter. Meanwhile the table below, adapted from Brobeck [57], provides a brief scheme of the possible factors involved.

3.4.4 *Ingestion of sodium chloride*

We have already seen that rats have the ability not only to adjust their food intake to energy needs but also to vary their choice of foods to meet other bodily demands (§ 3.3.1). The content of inorganic salts, or

1. Taking food to cover. (*a*) (*top*) A large male wild rat (about 450 g) drags a lump of horse liver of about his own weight towards his nest. (From Barnett & Spencer [31].) (*b*) (*centre*) Dragging a fowl's egg (*Gallus*) from a nest by the use of the forelimbs. (Netherlands Department of Agriculture.) **2.** (*bottom*) An experiment on food selection. A wild rat eats whole meal in preference to wheat grains or sugar. (From Barnett [18].)

3. Retrieving of young. A female wild rat carries a nestling home. (1/1250 sec.) This, and photographs 4-10, 12 and 13 were taken in large cages designed for the study of social behaviour [23].

4. Recognition by odour. A male wild rat investigates another male. A similar posture is sometimes adopted when a female is approached (figure 37, page 109).

of some vitamins, may determine the quantity consumed. To continue our critique of the concept of hunger, we can add that this has led some people to talk of 'special hungers', in addition to the 'general hunger' for energy-giving food. Clearly, this way of speaking can be accepted if it is first laid down what an expression such as 'sodium chloride hunger' is to mean: that is, what observable processes or events it is to refer to. But we must expect that, as with 'general hunger', the processes involved will be complex.

FACTORS INFLUENCING FOOD INTAKE

initiate or increase feeding	'reflexes'	stop or decrease feeding
lowered supply of glucose	sense organs (olfactory, gustatory, etc.)	exercise
		dehydration
decline in body temperature	sensory neurons	rise in body temperature
cold environment	brain stem 'centres'	warm environment
contractions of empty stomach	motor neurons	stimuli from mouth and pharynx
	muscles	distension of stomach

The internal processes involved in the selection of special dietary components have been studied for sodium chloride, and a recent report by Epstein & Stellar has shown something of the means by which salt ingestion is related to need. In general, there are two kinds of process which could be involved. First, it might be found that preference is shown for salt, or for salt in solution, only when the taste or some other cue had come to be *associated* with a particular change in the state of the body. This would then be an instance of learned behaviour of a familiar sort – one already exemplified in this chapter. This requires that the substance should have a fairly rapid (presumably beneficial) effect. Second, it is possible that a deficiency of salt could lead to an immediate change in behaviour, dependent perhaps on a change in the response to the specific taste of NaCl. This would be more difficult to prove conclusively; it would in any case be expected only when (i) the substance had a distinctive taste, and (ii) when it was present in food or drink in amounts sufficient for the taste to be effective. This second requirement might well not apply, say, to vitamin B_1 (thiamin).

Epstein & Stellar have brought evidence for both these possibilities as far as sodium chloride is concerned. They used adrenalectomized rats; these, as we saw above, not only need more salt than normal rats but can also select a diet to meet this need. Some of the rats were allowed access to salt only after a severe state of deprivation had been reached. These rats began drinking large amounts of salt solution at once; evidently post-operative experience was not essential for the development of behaviour adapted to the need.

An important question is whether such behaviour is related to a change, perhaps resulting from a lowered blood salt level, in the response of the taste organs to salt. Epstein & Stellar used an ingenious method to investigate the influence of taste. They administered ion exchange resins in such a way that rats drank, and tasted, a three per cent NaCl solution, but actually absorbed through the gut wall only as much salt as would have come from a 1·5 per cent solution; the rest of the salt was taken up by the resins. In these conditions the rats drank as much salt as they would have taken had the solution itself contained only 1·5 per cent NaCl: that is, they drank the amount appropriate to bodily need; the effect of the internal environment was thus greater than that of taste [93]. Pfaffmann has found, correspondingly, that the response to NaCl and other substances of the sensory nerve fibres from the tongue, recorded electrically, does not alter when internal bodily changes alter behaviour towards the substances [243].

However, as would be expected, in some situations learning dependent on taste is important. Adrenalectomized rats were allowed access to three per cent NaCl immediately after the operation, and not after a delay while deficiency developed; they then only gradually overcame their usual aversion to the salt solution; however, they eventually adopted the physiologically necessary habit of drinking it in large amounts. In this gradual change of behaviour, it may be suspected, the taste of the salt acted as a 'cue' or stimulus which came to be associated with its internal effects; the salt thus constituted a reward [93].

These observations provide only the beginnings of an analysis of the control of salt intake. They show once again that a rat may be responding to more than one kind of stimulus during behaviour adjusted to a specific need. The main categories of stimuli are (i) those that arise from the need itself and (ii) those that have no necessary causal connexion with the physiological state of the animal. The second group are, at least superficially, the more readily understood: they are the stimuli to which the animal *learns* to respond because of their association with the attainment of a consummatory state. The first group may or may not be involved in a learning process. They are more obscure and difficult to study, because the receptor organs involved are internal, and rather little is known about them. Accordingly, study of the physiology of selection of other special dietary components, notably the water-soluble vitamins, is exceedingly difficult, and so far has hardly been attempted.

The general interest of the work on salt ingestion is that it illustrates further how feeding behaviour may, in varying circumstances, be influenced by a diversity of 'motives'. The regulation of total food intake, of the intake of sodium chloride and of the ingestion of other dietary components – each of these has its internal mechanisms, partly independent but sometimes, as in the case of food and water, obviously interacting [306].

3.4.5 *The central nervous system*

Granted that eating and drinking are regulated by internal states such as the blood content of metabolites or the tonicity of body fluids, there remains the question of the apparatus which responds to these peripheral changes. Under this head we shall describe the special relationship of one part of the brain, the hypothalamus (figure 58, page 175), with the regulation of food intake.

The hypothalamus may be regarded as the most anterior part of the autonomic, or visceral motor, nervous system and also of the less well described visceral sensory system. It has complex connexions, both afferent and efferent, with the cerebral hemispheres and the thalamus. It is the most vascular part of the brain, and it is possible that it responds directly to changes in the composition of the blood. Injection of exceedingly small amounts of sodium chloride solution into the hypothalamus has made animals begin drinking when they had previously refused water. Electrical stimulation of parts of the anterior hypothalamus may influence the functioning of the gut, for instance its peristalsis; stimulation of posterior regions can alter the blood pressure and other aspects of the internal state of the body. The anterior is thus especially related to parasympathetic function, the posterior to sympathetic.

The hypothalamus is not regarded as a 'reflex centre': the visceral sensory tracts end lower down in the brain stem, and the visceral motor tracts which pass to motor neurons in the spinal cord also take origin more posteriorly. The hypothalamus has the function, evidently, of activating or inhibiting reflexes. But it does more than this. Local stimulation may set going complex patterns of behaviour: one example is fighting, which is associated with sympathetic stimulation; another is the whole sequence of activities involved in feeding (parasympathetic). Evidently, the motor output of the brain is to an important extent organized in the hypothalamus, especially that concerned with homeostatic and other visceral functions and with the performance of stereotyped action patterns.

The preceding statements apply quite generally to mammals, and are based on experiments on several species. Work on rats enables us to go into more detail on the regulation of feeding. Since 1939 several groups of experimenters have studied the disturbances of feeding brought about by damage to the hypothalamus. Bruce & Kennedy injured the tuberal area of the hypothalamus on both sides and so produced exceedingly obese rats; the increase in body weight was due to excessive eating. There was a constant relationship between the energy value of the diet and the increase in body weight, regardless of the composition of the diet: that is, high fat or high protein diets failed to alter the effects of the lesions. In the same way, control animals kept a constant weight despite changes in the composition of the food offered. Food intake (as we have seen) is 'calorimetric'; and these experiments show that the *cessation* of eating depends on the functioning of specific groups of hypothalamic neurons [59]. However, as Kennedy had already found, the voraciousness, or hyperphagia, of rats with these lesions is only a temporary phenomenon: after some weeks the rat, having attained a remarkable fatness, resumes a nearly normal intake without reverting to a normal body weight [169].

Just as the cessation of eating depends on particular groups of neurons, so does beginning to eat. Anand & Brobeck made minute injuries in different regions of the hypothalamus. Damage to the ventromedial nuclei produced the over-eating and obesity already described. At the same level, in the extreme lateral part of the lateral hypothalamus, they found a region of which destruction on both sides led to a complete failure to eat. Evidently, the very small groups of cells concerned have quite specific functions, since damage on a similar scale in regions near the nuclei mentioned does not have the same effect. If both the ventro-medial and the lateral nuclei are destroyed, the result is the same as destruction of the lateral centres alone: the animal stops feeding. This suggests that the ventro-medial group acts by inhibiting the lateral 'feeding' centres'; but this is not certain: it may be that both regions of the hypothalamus act directly on more posterior centres in the brain stem [6].

Many experiments illustrate the difficulty of devising methods to cope with the fine detail of neural function. Morrison & Mayer made minute lesions in the median eminence of the hypothalamus of rats, and found that adipsia was the main effect; but aphagia also developed. Further, the regions in which lesions produced the greatest effect on drinking also had a maximum influence on eating [227, 228]. Another problem is that brain lesions which lead to behavioural or sensory defects are often followed by a gradual recovery of function. This is true, although there is no regeneration of brain tissue. Such recovery has been demonstrated in rats in which hypothalamic damage had inhibited eating. Teitelbaum & Stellar kept the rats alive by feeding them through a stomach tube, and found that feeding behaviour was restored in six to sixty-five days. The rats regained and then held their original body weights. Their feeding behaviour was changed: for instance, unlike control rats, they ate much corn oil when this was offered; the recovered rats were also slower than controls to adjust their calorie intake to an increase in the non-nutritive bulk in the diet [308].

We saw above that feeding and increased general activity go together. Brobeck has reviewed the relationship between the effects of hypothalamic lesions which influence feeding and those that alter activity. But the facts so far available fail to make a coherent pattern. In general, hyperphagia due to hypothalamic damage is accompanied by excessive activity: the appetitive or variable component of the behaviour increases, as well as the time spent on feeding. However, it is possible to make lesions in the front part of the hypothalamus which induce excessive activity but no disturbance of feeding. What is worse, injuries to the tuberal part of the hypothalamus may be followed by reduced activity with *or without* a lowered food intake. There is in fact no necessary correlation between locomotory behaviour and the altered feeding resulting from these operations [57].

The most obvious comment on these facts about the hypothalamus is that they expose how little we know about central nervous function. The methods used for making lesions or for injecting substances represent a notable technical achievement: the instruments used are ingenious, delicate and finely made, and their use requires considerable skill. Yet the effects they produce are crude and gross compared with the fine structure of the brain as we see it, uncomprehendingly, under the higher powers of the light microscope. As we shall further find in § 9, we as yet lack even a language in which a lucid account of brain physiology can be given: we speak of 'centres' which 'control', 'organize' or 'coördinate' behaviour, but these terms lack the precision we would like them to have.

We can see that in the hypothalamus there are structurally separate arrangements concerned respectively with starting feeding and with stopping it. The study both of overt behaviour and of the peripheral internal factors which influence feeding has shown that the beginning and ending of a period of feeding have to be considered separately. One way of saying the same thing is to assert, as Brobeck does, that appetite and satiety are distinct phenomena [57]. Thus to this extent the facts from the different levels of investigation are in accord. One may further suspect that the contiguity, within the hypothalamus, of nuclei particularly concerned with general locomotor activity and many other aspects of bodily function is connected with the close relationship observed between feeding and other action patterns. Among these is drinking, which is certainly 'controlled' by the hypothalamus; and food and water intake are intimately linked. Again, the hypothalamus is concerned with the regulation of body temperature in response to changes in the environment; and we have seen how food intake too is affected by the temperature of the environment. Nevertheless, despite all this, we have no coherent account of the functioning of the hypothalamus, nor is one yet in sight.

3.5 CONCLUSIONS

The distinctive feature of the diet of rats is its variety: instead of specializing in a particular kind of food, rats have specialized in omnivory. Accordingly they have a system of behaviour which enables them not only to find food in a substantial area around their nest but also, to some extent, to select food according to need.

The exploratory behaviour of rats keeps them informed on food supplies in their neighbourhood; superimposed on it is an additional tendency to be active in hunger (or other deprivation). All possible food materials are sampled, though some substances with (to man) strong tastes or odours are rejected. There are also certain fixed preferences, for instance for sweet substances.

These preferences, seemingly independent of experience, have no evident homeostatic function; in detail they differ between wild and laboratory rats. The primary homeostatic effect is the maintenance of a steady energy intake. The selection of favourable foods, and avoidance of unfavourable (toxic) mixtures, both depend a good deal on learning: foods are sought or avoided according to their previously experienced internal effects, at least if these effects are fairly rapid.

Rats also display a number of behaviour patterns related to feeding, for instance hoarding. These give further examples of the way in which a complex activity depends on an intricate interaction of fixed capabilities with the effects of experience.

The study of feeding in rats provides the most detailed example we have of the physiological analysis of 'drive' or 'motivation' (§ 8). A good deal is known of the internal processes which determine when eating and drinking begin, and also what makes them stop; something is also known of the physiology of the selection of different concentrations of sodium chloride. The factors controlling ingestion are both peripheral and central: peripheral effects include the composition of blood and other body fluids and also distention of the stomach; centrally, certain parts of the hypothalamus occupy key positions in the 'control' of ingestion.

It is not known, however, what internal system provides the array of negative feedbacks which ensure that an animal maintains a steady, or a regularly increasing, body weight. It may be that there are alternative systems which operate in different circumstances or at different stages in an animal's life history. Blood glucose concentrations and the magnitude of adipose tissue depots may be involved. The plurality of the internal processes which control feeding is reflected in the various indices of 'hunger'. For instance the readiness with which a rat will press a lever to get food does not vary in parallel with its willingness to accept a distasteful mixture. The term 'hunger', whether used to refer to overt behaviour or to internal processes, is therefore often misleading, unless the exact behaviour or processes referred to are specified.

4

Rat 'Societies'

Even among beasts of prey the bloody wolves, who found
some selfish betterment from their hunting in packs,
had thereby learn'd submission to a controlling will,
their leader being so far charioteer of their rage;
while pastoral animals, or ever a drover came
to pen them for his profit, had in self-defence
herded together; and on the wild prairies are seen
when threaten'd by attack, congregating their young
within their midst for safety, and then serrying their ranks
in a front line compact to face the dreaded foe.
　　And this parental instinct, tho' it own cousinship
with Breed, was born of Selfhood. A nursing mammal,
since she must feel her suckling a piece of herself,
wil self-preserve and shelter it as herself; and oft
'tis hard to wean.

ROBERT BRIDGES

4.1 THE DEFINITION OF SOCIAL BEHAVIOUR

In this chapter we turn to activities that ensure reproductive success
rather than individual survival. Like all mammals, rats form family
groups consisting of a female and her young; such groups are necessarily
bound together during the period of dependence of the young on their
mother. But rats, wild or tame, are also social animals in the sense that
the adults too collect in groups whenever they have the opportunity.
Wild rats live in colonies which have nesting sites and feeding grounds
in common; there are evidently powerful influences which make them
assemble together. At the same time there are disruptive elements in
their behaviour towards members of their own species: in certain circumstances wild rats fight vigorously among themselves.

　　This chapter, then, is on *social behaviour*. In this book this term refers
to all behaviour which influences, or is influenced by, other members of
the same species. The term thus covers all sexual and reproductive
activities, all behaviour which tends to bring individuals together and
also all intraspecific conflict. The preceding sentences constitute a
stipulative definition. For reasons further discussed in Appendix 1, it
follows that the statements they contain cannot be 'correct' or 'incorrect': they express an intention to use the world 'social' in a particular

way *in this book*. The word is so used on the grounds of *convenience*.

Social behaviour so defined covers a great variety of complex relationships. As far as rats are concerned these are often best studied in groups of wild individuals, either in conditions approaching the 'natural' or in contrived situations which nevertheless permit the rats plenty of scope for varied behaviour. The importance of these requirements is shown in the following statement by Munn: 'Relatively little research has been done on social behaviour in rats primarily because rats are not especially influenced by each other's actions' [233, p. 465]. We shall see below that the second part of this statement is mistaken. The fact that it could be made, in an admirable work written by an authority on 'rat psychology', is due to the concentration of psychological research on tame rats in conditions which leave them little scope for social interactions. Until recently, there has been little reliable information on what wild rats do among themselves in their colonies, let alone any explanation of their activities in terms of the stimuli that evoke them. This gap has now been partly filled, and a good deal of §§ 4.2 and 4.3 consists of *descriptions* of social behaviour in wild rats. These enable us to give a fairly full answer to the question: just what kinds of relationship hold between members of the species, *Rattus norvegicus*? Something will also be said, in § 4.5, about *R. rattus*. At the same time, partial answers will be given to the following questions: (i) what are the conditions in which the different types of behaviour are displayed? (ii) what are the specific stimuli involved? and (iii) what is the survival value of the various kinds of behaviour? Unless other sources are quoted, the statements in the next three sections are based on a study by Barnett [19]. A few of the details have not been previously published.

4.2   THE GROUP

4.2.1 *The causes of cohesion*

4.2.1.1 *Non-social effects*. Before describing the social releasing stimuli which make rats assemble, something must be said of other inducements to draw together. Members of one species often associate because they all respond in the same way to some feature of their environment such as a source of food. This is not social behaviour as defined, unless the individuals concerned also attract each other directly in some way. Sometimes rats may be brought together by the amenities offered by a particular area. To be suitable for colonization by rats a biotope must offer food, cover and nesting sites and material. Cover and nesting sites may be (and away from buildings usually are) provided by earth burrows. However, we shall see below (§ 4.3) that there are severe restrictions on the extent to which strange rats can come together to form a stable group. Nevertheless, it is, on existing evidence, possible that immature rats of both sexes, and adult females, sometimes join with other groups as a result in the first place of the suitability of a biotope such as a large refuse dump, a glue factory, a food warehouse or a farm; all such places, in the absence of strenuous action by the human occupiers, are liable to harbour large colonies of rats.

The associations which rats form may then be in part a result of nonsocial stimuli. Nevertheless, they are to an important extent dependent on social interactions. This has long been assumed to be so, and expressions such as 'gregarious instinct' have been used to name either the behaviour involved or a hypothetical inner cause of the behaviour. Such ways of speaking have not proved very helpful in analysing behaviour. We need to know whether animals tend to approach, rather than avoid or ignore, each other; and, if so, in what circumstances they do so and what are the exact stimuli responsible. Some early experiments with laboratory rats, reviewed by Munn [233, p. 465], failed to demonstrate any tendency to assemble together. One method was to put a rat in an apparatus in which it had the opportunity to cross a barrier to reach another rat; in another investigation rats were offered a choice between a compartment containing both food and other rats, and one containing only food. It is not surprising that the only positive conclusions reached in these inquiries was that rats possess a strong 'exploratory drive'.

4.2.1.2 *Huddling.* Fortunately, Soulairac has more recently given a formal demonstration of the familiar fact that laboratory rats do associate and huddle together in groups. He used a relatively large cage and allowed a group of four males and four females to live in it undisturbed. In these conditions, although there was opportunity to separate off into isolated individuals or pairs, in fact the rats slept together in a group. Only parturient females isolated themselves [295]. Similar observations have been made on wild rats in large cages. If adult rats, strangers to each other and unfamiliar with the cage, are placed it it, they explore in the manner described in § 2.2.1. But they also tend, even at this early stage to group together (figure 22), especially in the corners of the main cage. Later, they can be studied during periods when they are inactive; these occur mainly during the day. They then sleep or rest in groups, either in the nest boxes attached to the main cage or in the corners of the main cage as before. This behaviour is not merely a product of the shelter or the tactile stimuli offered by these places: if rats do rest on the floor of the cage they usually do so all in one corner; the corner chosen may vary, but the grouping remains rather constant. Similarly, when several nest boxes are available, only one or two may be occupied; when more than one, each is (as a rule) used by a group. Again, the choice of nest box may vary over a period, even though conditions in and around the cage remain unchanged. These statements apply, not only to colonies of *R. norvegicus* alone but also to groups of this species and *R. rattus* together.

22. A group of wild rats. These rats had recently been put in a large, strange cage; they were strangers to each other. In the intervals of exploration they collected in groups of the kind shown. Three are *Rattus norvegicus*; the darkest one is *R. rattus* (black variety). (After Barnett [19].)

Rats often sleep, not merely lying alongside each other but piled in a heap. From time to time one at the top wakes or rolls off and then burrows its way into the bottom. This has often been observed in young wild rats [26]. *Huddling* is one of a number of ways in which rats derive cutaneous stimulation from each other. What is its biological significance? It is commonly thought that huddling has a heat-conserving function. If this is its function, the stimuli to which a rat responds may be the temperature of the skin of other rats, the pressure of other rats' bodies or something else less obvious. It is certain that, in some circumstances, the heat provided by other animals is important for small mammals. This has been shown for rats by Gelineo & Gelineo [110] and for mice by Prychodko [247]. Nestling rats especially are dependent on heat from their mother, since they are unable to regulate their own temperature until they are at least eighteen days old. The colder the environment, the more important an external source of heat becomes. Nevertheless, Benedict has shown that in ordinary conditions in the laboratory huddling has no energy-conserving effect [44].

Yet rats continue to huddle in these conditions. Huddling gives the impression of being an example of 'innate' behaviour but nothing is known of how it develops. The earliest contacts a rat has with other rats are with its mother and litter-mates in the nest immediately after birth; the first environment of a rat is in fact one in which cutaneous stimulation from other rats is almost continuous. Whether this influences later huddling has not been investigated. Whatever its mode of development, its universality suggests strongly that it has an important function, and this assumption is given some support by our knowledge of other ways in which rats stimulate each others' skin senses.

23. Amicable behaviour. A *Rattus rattus* 'crawls under' a *R. norvegicus*. (After Barnett [19].)

4.2.1.3 *Cutaneous stimuli*. Wild rats display with some regularity four types of behaviour, besides huddling, which involve mutual cutaneous stimulation. The most distinctive is *crawling under*. This act, which is accurately described by its name, is illustrated in figure 23. It is performed by both *R. norvegicus* and *R. rattus*, and has been observed especially among males in situations which might lead to conflict. For instance, if a strange male approaches a non-aggressive male in the latter's territory, the resident may crawl under the stranger. A stranger may also crawl under a resident. Crawling under is often seen among young rats, during play. A second, and allied type of act may be called *walking over*. This expression, again accurately descriptive, refers to an act which no doubt often takes place incidentally, but which certainly is also carried out as a specific response to the presence of another rat. Observation shows that rats often literally go out of their way to crawl under or walk over other members of their colony. A third type of contact behaviour is mutual

24. Amicable behaviour. Grooming of one wild rat by another. The left-hand male was a newcomer, and was attacked as well as groomed. (After Barnett [19].)

grooming. This consists of a gentle nibbling of the fur on any accessible area of skin. Like crawling under, it is seen mainly in conflict situations (figure 24), but it is not confined to them. Fourth, and last of the distinctive acts under this heading, is *nosing*, in which one rat gently pushes with its nose at the flank, usually near the neck, of another rat (figure 25). This is seen in a variety of situations, including that in which a female in oestrus approaches a male (§ 5.2.1.1.)

4.2.1.4 *Odours.* It will be shown further below (§ 4.3.3) that contact stimuli probably help to determine whether a rat attacks another rat. This undoubtedly applies still more to olfactory stimuli; but these are more difficult to study and we know little about them. The importance of the olfactory sense for most mammals may be inferred from a superficial study of their behaviour and of the relations of the olfactory organs to the brain. The part smell plays in feeding and exploration has already been mentioned. But rats not only sniff all food and similar materials they encounter; they also sniff other rats, especially strangers and potential mates (figure 38, page 110; plate 4). Most mammals have an elaborate equipment of glands for secreting substances which stimulate other members of their own species. Le Magnen has shown that male laboratory rats can distinguish the odour of a female from that of a male, and of a female in oestrus from other females [205].

25. Amicable behaviour. A male paws and nuzzles a female; the latter adopts a 'submissive' attitude.

Rats also leave odour trails during their movements about the living space. Where they run regularly over a light surface they leave a dark smear. Reiff has shown that urine and genital secretions contribute to the odour trails, and that the trails are followed by other rats [251]. In artificial colonies the trails always have an attractive effect, even when the rat is a newcomer to the enclosure and to the colony. Hediger, in his excellent study of wild animals in captivity [139], suggests that the scent marks left by many mammals at fixed points in their territory are deterrents to other members of their species, comparable to bird song and other displays; but it is uncertain whether this is ever their function in fact. It is evidently not so for wild rats. Perhaps the marks have a mnemonic function: they may help an animal to remember its way about. The importance of this has already been mentioned in § 2.4.1. Scent marks or trails may help all members of a colony in common [252].

4.2.1.5 *Noises.* The squealing or squeaking of rats, though easily heard, also lacks detailed study as yet. Its social effects are evidently of several kinds. First, nestlings, especially before they become active, squeak a great deal, in chorus. From the work of Wiesner & Sheard on laboratory rats it seems likely that they make most noise when their stomachs are empty or when their skin temperature falls below a certain level. The second is of especial importance when a young rat has strayed or fallen from the nest: the noise is then a signal to the mother to retrieve it [330].

Second is a gentle piping or whistling made by adults. Males sometimes make this noise when approached by another male, especially if the latter has adopted the 'threat posture' described in § 4.3.3 below; it is heard from males defending a nest against a visitor, and in these conditions it is sometimes effective without any further action on the part of the defender. Females, too, make this sort of noise, when defending a nest containing young; they also make it when approached by a strange male. It is easy to make a plausible suggestion about the function of this sort of noise: like the cutaneous stimuli described above (§ 4.2.1.4) it presumably tends to prevent harmful conflict.

A third kind of noise is louder and more complex. This is the squeal known to anyone who has handled or attacked wild rats. Its social significance is of two kinds. First, on hearing it, other rats take cover. It is, in fact, an alarm call of the sort familiar in many other species. To the unaided human ear *Rattus norvegicus* and *R. rattus* have very similar alarm calls; and this is in accordance with expectation since as Marler has pointed out, alarm calls of different species of birds are similar [209]: they are difficult to locate, and there is no advantage in their being species-characteristic as there is in the case, for instance, of a sexual signal. The second social effect is in defence. If a rat is defending a nest against a visitor, and a mild whistle has no effect, the noise intensifies and becomes similar to – perhaps identical with – the alarm call. This noise probably also has a non-social function, in that it tends to deter predators: it is certainly disconcerting to man.

Another noise, tooth-chattering, made by male rats, is dealt with below (§ 4.3.3); and in § 7.3.1 reference is made to the use of sounds above the limit of human hearing.

4.2.2 *The Family*

The preceding account has shown that in aggregations of adult rats the individuals perform various stereotyped acts which may be called 'amicable': they are directed at other rats and they are the opposite of aggressive. Their importance will be further shown in the account of fighting in the next section. One question that may be asked is: how does amicable or, more generally, gregarious, behaviour develop? To what extent is it fixed in the individual, regardless of experience?

These questions cannot yet be answered; but something can be said about the early social experiences of rats.

In all mammals the foundation of group relationships must be sought in the family: in particular, in the relationships between mother and young and among litter-mates in polytocous species. In rats the male plays no part in the care of young: he ignores them. The behaviour of the female towards her young has been described in detail by Wiesner & Sheard [330] and reviewed by Munn [233]. The first activity of the female directed towards her young occurs, as a rule, before parturition is completed: she adopts a 'head between heels' position, and licks the infant as it emerges; when the infant is finally born the membranes are neatly stripped off, the umbilical cord bitten through and the placenta eaten. The infant is also further licked, especially in the genital region. For a long time this licking was thought to have the sole function of cleaning, but recent experiments have shown that it is a remarkable and special case of the importance of cutaneous stimulation in the development of behaviour. This was discovered when Reyniers & Ervin tried to rear rats from birth artificially. The young accepted milk from the 'bottle', but they failed to release their urine and faeces in the normal way and so died. This failure was found to be preventible by gentle stimulation of the perineal region. Evidently the reflexes involved require this stimulation before they become effective [253].

Once the immediately post-partum operations have been carried out, the female displays, for a substantial period, further stereotyped responses which contribute to the survival of the young: these include adopting the 'nursing posture' which allows the young to suck, retrieving young which have strayed from the nest (plate 3), more licking, nest-building and defending the nest. All these activities are the same in both wild and tame rats, except that defence of the nest may not be performed by the latter. The young, for their part, move by wriggling their bodies and pushing with their limbs; they turn their heads towards a nipple and suck; and they squeak; eventually, at about sixteen days, their eyes open and they become active. Seitz has studied the effect of litter size on the behaviour of laboratory rats: those reared in litter groups of six were less wild, less 'anxious', more exploratory and hoarded less than others reared in groups of twelve [281]. Perhaps this was because they had more milk [191, 192]. Little else is known about the mechanisms of the development of behaviour in infant rats; there is special need for work on the effects on behaviour of social deprivation (§ 8.4.2).

5. 'Fighting'. During an experiment on social behaviour in wild rats a male newcomer takes to flight and is pursued and bitten by a resident.

6. 'Fighting'. (*a*) (*centre*) A male wild rat, though unwounded, collapses under attack. (*b*) (*bottom*) The attacker sniffs the now motionless intruder; the latter no longer presents the constellation of stimuli which evoke attack.

7. 'Submissive' posture (*top*). In an experiment on interspecific relations, a male *Rattus norvegicus* newcomer adopts a non-aggressive attitude on the approach of a resident male *R. rattus*. (From Barnett [19].) **8.** Interspecific conflict. (*a*) (*centre*) A momentary pause in a fight between a light-coloured *Rattus rattus* (*left*) and a *R. norvegicus*. (*b*) (*bottom*) A resident *R. norvegicus* makes a severe attack on a strange *R. rattus*. (Both 1/1250 sec.) (From Barnett [19].)

4.3 CONFLICT AND TERRITORY

4.3.1 *Evidence from ecology*

Although 'amicable' (including parental) behaviour is of such obvious importance, much more attention has been paid to conflict and aggression, not only in wild rats but often in other species. Perhaps this reflects the preoccupations of experimenters and observers. The first accurate information on rats came from studies of populations; these in turn arose in part from the practical importance, for pest control, of a knowledge of the dynamics of rat numbers.

When most members of a settled rat population have been killed, the remainder breed more quickly and numbers rise again. At first the growth rate of the population is high, but later it declines, sometimes quite steadily, until a maximum is reached. Thereafter, if the environment remains unchanged so, probably, does the size of the rat population. The environment includes predation by men, cats, dogs, hawks and so on. A question both of theoretical and practical importance is: what determines the maximum? Where a fairly steady population size is maintained, this is usually, if not always, due to the operation of factors which act with increasing intensity as the population enlarges. Such agencies (reviewed by Lack [182]) are said to be *density-dependent*. This notion is not as simple as it seems at first sight, but it provides a convenient frame of reference for talking about rat (and many other) populations.

The most obvious factor which might act in this fashion is food. Consider a habitat in which the food supply is constant: for example, in an area in which rats are feeding on garbage, the daily rate at which garbage becomes available might vary little. If such an area were colonized by a few rats, their numbers would increase slowly at first, then more rapidly, until, eventually, it became increasingly difficult for individuals to find enough food; mortality would then increase and, as Davis has shown, pregnancy rates would decrease [83]. The population would level off. Other potentially density-dependent factors are the available nesting places (especially important for the rearing of young), predation and infectious disease. Finally, there is conflict with members of the same species, or intraspecific 'competition'.

Calhoun added strange rats of both sexes to established urban populations of *R. norvegicus*. Retrapping over a period of one to three months after the initial release showed that only sixteen per cent of the foreign rats remained: the rest had either died or emigrated. The indigenous rats kept to their usual, quite small, home range, whereas the foreigners moved widely [63]. A second study by Calhoun was of a single large colony (figure 26). The rats had different degrees of access to the one source of food, and before the population had reached its maximum there was evidence of conflict, though not specifically for the food. Rats living nearer the food molested those living further away when the latter were moving between the feeding point and their nests.

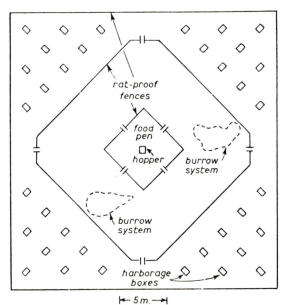

26. Experiment on social behaviour in *Rattus norvegicus*. The only food available was in the centre of the enclosure. Rats living in the peripheral areas were attacked on their way to food and had a higher mortality and were less fertile – an example of territorial behaviour and its effects. (After Calhoun [64].)

There were differences in growth and fertility which reflected the amount of persecution to which different groups of rats were subjected. Evidently rats defend an area around their nests against other members of their own species; that is, they are territorial animals, as further discussed below (§§ 4.3.4, 4.5). Some further support for this view is provided by the studies of Schein & Orgain on rats feeding on garbage. These workers established that typical rat populations in an urban environment had access to a food supply much in excess of their needs. They suggest that the distribution of the food in the biotope is important [269]. If the sort of effect described by Calhoun is general, then rat populations may sometimes be kept down by intraspecific conflict interacting in a complex way with the distribution of food and shelter [238].

That populations of rats may in fact increase at a declining rate to an asymptote has been shown in a number of investigations: urban populations have been reported on by Emlen and his colleagues [92], rural populations by Barnett and his colleagues [28], and sewer rats by Barnett & Bathard [27]. Nevertheless, the model of a closed area with constant amenities and a ceiling population can be misleading. One important qualification is that the area occupied by rats often increases with a rising population: further, as Barnett and his colleagues found in two villages in an agricultural area, new ground may be colonized *before* the maximum density is reached. The extent to which this is due to exploratory behaviour alone is not known but, although exploration must play a part, it seems likely that conflict also contributes to rat dispersal.

4.3.2 *Conditions for conflict*

We now turn again to direct studies of behaviour [19]. If one wishes to establish a stable, healthy colony of wild rats in artificial conditions, the best method is to begin with a group of sexually immature individuals. They need not be litter mates. Or a single adult male with one or more females may be used. In either case conflict is exceedingly unlikely; there are instances of males attacking females, but they are rare.

By contrast, there is one type of situation in which fighting is highly *probable*. This is when an adult male enters a region in which another adult is already established. This holds even if the resident male or males have been brought up from birth only with their mother and siblings, and consequently have never fought before, except in play. A series of experiments was carried out with small artificial colonies, some consisting only of adult males, some of males and females, some of rats which had been brought up together from the age of six weeks or younger. Of twenty males added to such colonies, eighteen died. There was no corresponding mortality among the residents. Similar observations, some unpublished, have also been made on the effect of introducing males into a large cage containing only one resident male [29]. By contrast, the addition of females rarely resulted in injury or death. No attacks on young rats were seen.

The attacks by residents on strangers are typical examples of territorial behaviour. The conventional definition of a 'territory' is 'a defended area'. In wild rats territorial behaviour is, except in one type of situation, confined to the males. The exception is the behaviour of a female with young nestlings; the defence in this case is not of an area coinciding with, or similar to, the home range, but only of the nest itself: many lactating females defend their nests against all visitors, regardless of age, sex or familiarity. However, the intensity of nest defence varies greatly: some females allow adult members of their colony, of both sexes, to share their nest with the young. The two kinds of territory are common in mammals. In rats they are particularly distinct, since females take no part in defending the home range and males do not defend a nest except when they are cornered in one by a more powerful male.

The account so far given does not, however, prove that fighting between males is purely, or even mainly, territorial. It might be that males sometimes fight each other 'spontaneously', or for a female. It has already been stated that males brought up together from youth do not fight; but even adult males, strangers to each other, put together in an artificial colony may live tranquilly together. For this to happen, two conditions must be observed: first, they must all be strange to the cage or enclosure; second, they must all be introduced at the same time. Even an interval of ten minutes between one male and the next may alter the situation. Provided these conditions are satisfied, even all-male colonies of *Rattus norvegicus* and *R. rattus* together may be maintained without conflict. Hence an encounter between males need not involve fighting; indeed, in a settled colony, a clash between males is unusual: the rats either ignore each other or behave in one of the 'amicable' ways already described. These observations show that crowding is not by itself a cause of conflict among wild rats.

There is next the question whether male rats fight for females. At first, during the study of experimental colonies, it seemed certain that they did. Colonies consisting of both males and females were compared with others containing only males, and with mated pairs in small cages. As already mentioned, the all-male colonies were often entirely peaceful: in some there was no mortality over periods of many weeks, and the rats grew well (figure 27). The same applied to the mated pairs. In the mixed colonies, by contrast, deaths among the males were the rule: in two extreme instances all except one died. There were very few deaths among the females, and there was no evidence that these were a result of fighting. Despite these facts, direct observation of the behaviour of the rats showed that there was *no fighting for females at all*. (This is in contrast to the situation in 'higher' mammals, for instance the baboons, *Papio hamadryas*, studied by Zuckerman [344].) When a female is in oestrus, one or more males concentrate on copulating with her; there is no competition, and the female accepts any male indiscriminately. When a female is not in oestrus she is usually ignored; but, during the period preceding oestrus, males sometimes sniff her genitals or attempt coitus unsuccessfully. A possible explanation for the fact that, despite this, conflict, injury and death were all greater in the presence of females must be deferred to § 8.2.3.2.

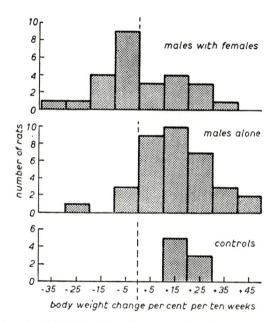

27. Growth and social relationships. Experimental colonies of wild rats included either males and females or males alone. The controls were males, each alone in a small cage with a female. There was severe conflict only in the male-female colonies (see text) and in them many males lost weight although food was available in excess. (After Barnett [19].)

We are consequently left with only a territorial situation as a certain cause of conflict. This situation requires that an adult male should be established in an area, and that he should encounter another adult male not a member of his own colony. It now falls to describe exactly how rats behave in these circumstances and, in so far as they are known, just what stimuli they are responding to.

4.3.3 Fighting and its consequences

4.3.3.1 *Motor components.* The behaviour that has been called 'fighting' in the preceding account is an example of a stereotyped behaviour pattern with a number of distinct components. In a typical, complete fight between a resident and a stranger, these components are all displayed by the resident; the attacked rat may adopt a defensive posture (figure 28), he may squeal or he may run away, but he does not fight back. In fact the term 'fighting' may quite reasonably be regarded as a misleading anthropomorphism: 'assault with battery' suggests more clearly what actually happens.

28. Fighting. Attacker (*right*) and attacked newcomer both momentarily in a defensive position. The attacker has raised hair. (After Barnett [19].)

The first activity performed by an aggressor is often *tooth-chattering*. This is not confined to *R. norvegicus:* it occurs in *R. rattus* and also in other genera of rodents such as *Cavia*. It goes on while the rat remains immobile, after he has detected the presence of an opponent, and while he approaches the latter. Its significance is not known. It might act as a deterrent preventing further approach by a stranger. Such signals, leading 'automatically' to the flight of a newcomer without injury to either party, certainly occur in many vertebrate species; but close observation of rats has provided no evidence that tooth-chattering acts in this way. It would, like any other stimulus, come to be *associated* with attack by a rat which had experienced it on a number of occasions, but there is no evidence that this sort of effect is important either. It is possible that it has no social significance: a male about to attack is in a highly specific physiological state; as he advances towards his opponent he urinates and defaecates and his hair is raised; these are signs of activation of the autonomic nervous system. Perhaps the tooth-chattering is only a by-product of this activation; certainly it ceases at the start of the more vigorous part of the proceedings.

In at least one species of the Murinae, the crested rat (*Lophiomys ibeanus*), the raising of the hairs when the animal is disturbed reveals a striking pattern of black and white [73, p. 219]. This perhaps has a deterrent effect on predators or rivals. In *Rattus*, however, there is no evidence that pilo-erection acts as a visual signal.

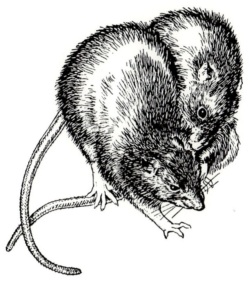

29. 'Threat posture'. Two males of equal status both adopt the arched-back attitude. No fight followed.

Typically the next stage is initiated by the adoption of the *threat posture* by the attacker (figure 29). The back is maximally arched, all four limbs are extended and the flank is turned towards the opponent. While in this attitude the rat may move round his victim with short, mincing steps, still presenting his flank. This too occurs in *R. rattus*, and similar behaviour has been described in mice, *Mus musculus*. It is likely that this act has a true social function. Sometimes, two males of equal status in a colony both adopt the threat posture and direct it at each other. The two rats may prance around for a few seconds, flank to flank and generally in contact (another example of cutaneous stimulation), and finally break away without further conflict. This behaviour suggests that the posture, which is highly distinctive, has a function like that of many displays and noises, namely, the prevention of injury in a situation which might otherwise result in harmful fighting.

The next stage, in both species, is that of *leaping* and *biting*. This can be followed in detail only by means of high speed filming. The attacker springs in the air and comes down on his opponent repeatedly striking at him by rapidly adducting his extended forelimbs (figure 30); he also bites, usually an ear or limb or the tail, less often the skin of face, rump or belly. Typically, a bite is exceedingly brief: as soon as the jaws close

30. Fighting. A fierce male attacks a strange male by leaping and biting; this movement is accompanied by rapid adductions of the forelimbs. (After Barnett [19].)

they open again and the rat leaps back. Rarely, very aggressive *norvegicus* males bite and hold on. A leap and bite is followed as a rule by a second or two of violent movement on the part of both rats, towards the opponent on the part of the attacker, evasive on the part of the other rat, until both become disorientated (figure 31). These brief bouts are

31. Fighting. Both combatants fall.

followed by longer intervals in which both rats adopt a defensive attitude, or possibly a stance as if boxing (figure 32). Functionally, biting is obviously a pain-causing stimulus likely to evoke flight in the animal attacked, if non-violent sign-stimuli have failed to do so.

32. The 'boxing' position.

In many fights there is a complete sequence of stereotyped actions, and this might suggest that the different motor components of the total pattern are related to each other in a specific way. However, tooth-chattering, threat-posturing, leaping and biting may each occur independently of the others: there is no fixed chain or 'hierarchical organization' of the components of fighting behaviour in wild rats.

4.3.3.2 *Releasing and suppressing stimuli.* Fighting behaviour, then, is decidedly variable. And individual rats differ greatly in the intensity with which they fight and in the extent to which they perform the separate motor patterns. Nevertheless, there are certain conditions in which typical fighting can be predicted with confidence. We may now examine more closely what these conditions are and what stimuli are involved.

First, we have seen that the actors must be adult males. We may infer that fighting is performed only when there is sufficient male hormone in the blood. Second, a profound difference is made by the familiarity of the surroundings: a rat, however large and vigorous, is unlikely to attack another when in a strange place.

Given the internal or 'physiological' state and the appropriate environmental conditions, it is still the case that a rat will attack only some other living things, while it will merely sniff or entirely ignore others. The exact constellation of stimuli which induce attack is not known. Observation of behaviour suggests that a male rat's attention is usually drawn to a stranger first by the sounds of its moving about or through seeing it. This seems obvious, but should not be considered so. This is what would ordinarily happen, in parallel circumstances in our own species, but it does not follow that it may be expected in others. A rat's vision is inferior to that of man, and its olfactory sense plays a much bigger part in influencing behaviour. Further it is possible that, even if one rat does hear another, the sounds to which it responds differ from those that we hear (§ 7.3.1.1).

When we turn to visual stimuli we have to consider two aspects, namely, form and movement. That movement is important can be seen when an attacked rat falls over and lies still: nearly always, the attacker pays little or no further attention until his victim begins to move again. As for form, rats may attack not only members of their own species but also other mammals of similar size and smaller (for instance, mice, *Mus musculus*), and small birds.

In view of the diversity of creatures which may be attacked, the problem that looms largest in intraspecific conflict is not why members of the same species are attacked, but why in so many instances they are left alone. This problem is best considered in relation to the olfactory stimuli that influence the behaviour.

As with most other mammals, the importance of the olfactory sense in rats can be guessed from observations of behaviour. When a rat begins to emerge from cover, for instance a burrow, it may pause with only its head out, evidently sniffing the air. This is observed especially when an unfamiliar object with an odour, such as a man or a strange rat, is nearby. The rat does not move its head as if using its eyes, but its nostrils do move. When outside it may stand on its hind legs and apparently sniff the air again. Objects of all kinds are closely sniffed, especially other rats.

The latter probably present two kinds of odour, both of social importance, with a possible third. First, there is a 'type odour', determined by age and sex, perhaps also by reproductive state if a female with young is distinguished from other females. We have already seen that the responses of an adult male to a young rat, an adult female not in oestrus, one in oestrus and an adult male are all different. Second, it seems certain, that in a settled group, there is a 'colony odour', which enables a male to distinguish a member of his own family or group from strange rats. When strangers have been introduced into a colony, there is a marked increase in the amount of 'recognition sniffing' among the residents. Third, it is possible that individual odours can be distinguished. Kalmus has shown that dogs can detect differences in odour between human beings [166], but the point has hardly been investigated in the rat. Since rats do not form marital pairs there is no behavioural evidence of individual recognition among adults. Beach & Jaynes have brought evidence of recognition of young by laboratory females, but this was probably through their possession of a distinctive nest odour [41].

Attack, then, depends on a number of factors. The 'background' includes the internal state of the animal and his experience of the immediate surroundings. The immediate factors probably include auditory, visual and olfactory stimuli. Occasional instances of exceptional ferocity apart, a male rat does not attack a member of his own species unless, first, the opponent is an adult male (with, presumably, a characteristic odour) and, second, the opponent lacks the colony odour of the attacker's own group. In addition to the 'social suppressors' constituted by auditory signals and so forth, certain olfactory qualities seem to prohibit attack. On our present, incomplete knowledge, it seems likely that the failure to attack an adult female is due to an inhibition shared by all members of the same species. The inhibition due to colony odour, on the other hand, is obviously confined to members of the colony and is learned during a quite short period.

This summary of the causes of attack is still incomplete, since it ignores the marked variation among individuals in aggressiveness. This and related topics, must be dealt with next.

4.3.3.3 *Dominance and subordination.* Much of this chapter is based on observations of rats in small, artificial colonies. By analogy with other species, and of birds, it might be expected that, when such colonies are stable, there would be a form of dominance hierarchy at least among the males. Such hierarchies have attracted much interest among students of social behaviour, perhaps because of analogies with our own species. In the simplest case there would be a dominant or top rat with prior access to females, food and nesting places; a second rat would give way only to the first, and so on. In fact, nothing quite like this is found. It has already been stated that rats brought up from before sexual maturity form a group without conflict or hierarchy. When adults are put together, and conflict occurs, the relationships between the males are determined (as in dominance hierarchies) by the results of the fighting. Three types of males may be distinguished. *Alphas* are always large by comparison with other members of the colony: in no instance has a rat been seen to defeat another *colony member* much larger than itself (though *newly introduced* rats of considerably greater weight are usually defeated sooner or later). Such males move about without hesitation or any attempt to take flight from other rats; if there is fighting, they initiate it.

Secondly, *omegas* are the result of defeat by one or more alphas. These rats flee at the approach of an alpha. In the confined colonies in which these observations were made, omegas, after a day or two of persecution, were marked by their slow movements, drooping posture and bedraggled appearance. They lost weight, and died if not removed. A third category is needed for rats which, after defeat, adapt themselves to an inferior rôle: they have been called *betas*. They endure defeat without severe shock and succeed in feeding with enough freedom to gain weight. Omegas and betas associate together without conflict: there is no hierarchy.

There was no uniformity in the numbers of alphas, betas and omegas in the small colonies studied, but in conditions of intense conflict there were fewer alphas and often only one.

In unconfined colonies it seems likely that the adult males vary in status from alpha to beta (in the terminology used here); any rat with an omega status would soon die or emigrate. The function of hierarchical systems is presumably to ensure that conflict does not reduce the chances of survival of a population, even though it may do so for particular individuals. In wild rats it seems that harmful conflict is prevented, for the most part, not by the setting up of a 'peck order' but by inhibition of conflict among members of a single family or similar group; the inhibition is evidently strengthened by the various types of 'amicable' behaviour. Attack is usually released only by the arrival of a strange male; and this conflict presumably has the function of promoting dispersal. The functions of territorial behaviour have been much disputed [147, 182], but for rats it seems likely that the dispersal it causes opens up new areas for colonization, and makes full use of the food and other amenities available in a given area.

4.4   LABORATORY RATS

The preceding account will perhaps be strange reading for those whose only experience has been with laboratory rats. Munn, in his excellent compilation, so often quoted in this book, relegates the account of social behaviour to the chapter on abnormal behaviour. Another psychologist of distinction has remarked that rats have 'very little social life' [317, p. 229].

A few authors, not in the main stream of 'rat psychology', have paid some attention to laboratory rats as social animals. Small, at the end of the nineteenth century, before rat psychology had properly begun, made some reference to social interactions, but his main interests were in learning and in the ontogeny of behaviour [291, 292]. Buytendijk, in 1931, tried to break away from what he regarded as the unnatural situations inflicted on rats in mazes; he used a spacious observation box stocked with a variety of objects and giving scope for social interaction. Unlike most of the conventional workers of his time he was led to put much emphasis on exploratory behaviour, but the conditions of his experiments did not facilitate the study of social behaviour: not surprisingly, at that time, he did not know what to look for [62]. In 1952 Soulairac published a short description (already quoted) of the effects on behaviour of keeping laboratory rats in a group in a large cage with several nest boxes. He observed their marked tendency to huddle, territorial behaviour by parturient females and examples of temporary dominance among the males; his rats, in these conditions, also became less amenable to handling [295].

In other studies of social behaviour laboratory rats have been put in highly contrived situations in order to evoke specific forms of behaviour. Seward, for instance, unlike Buytendijk, knew exactly what he was looking for: he sought explicitly to establish a dominance hierarchy by arranging bouts of fighting between pairs of males. No true hierarchy emerged; but Seward found that past experiences during fights had much effect on subsequent aggression or timidity [284].

None of these researches gives a clear picture of the social repertoire of laboratory rats or of their differences from wild rats. Barnett made a short study of laboratory rats in exactly the conditions in which he had already observed social behaviour in wild rats. There was a complete absence of certain of the stereotyped behaviour patterns found in wild rats: of these the most important were the threat posture, which is a usual feature of aggression, and the crawling under which is the most distinctive of amicable acts. 'Fighting' in general was mild and resembled the playful wrestling of immature or female wild rats, not that of adult males. Newcomers, albino or hooded, put in established albino colonies, were soon absorbed; none died. There was, however, some evidence of 'stress' in the interlopers, since they lost weight and had

enlarged adrenal glands [24]. The loss of weight at least, however,
may have been due merely to the disturbance of transfer to a new cage:
Steinberg & Watson have reported a failure of growth in laboratory
rats due to disturbance only [302].

We can then summarize the behavioural differences, due to selection
in laboratories, of our domestic rats from wild ones. First, some components of fighting and of amicable behaviour have been totally lost.
Clearly, for most experimenters, rats which do not fight among themselves are to be preferred, and such rats may well have been selected for
breeding. The fact that certain amicable actions are lost too is not so
readily explained and deserves further inquiry. Second, the remaining
'fighting' behaviour tends to be immature in character and is almost
harmless. Third, wildness and savageness towards man have been
almost wholly lost; and Karli has shown that the tendency to attack
smaller animals has been much reduced [167]. The extent to which
these changes are necessary concomitants of the first group is uncertain. Fourth, and finally, fear of new objects in a familiar environment has been lost.

There is still open a great field of inquiry in the genetics of the behaviour of different strains of laboratory and wild rats. Some aspects of this
topic are further discussed in § 5.3.1.

4.5 RELATIONSHIPS BETWEEN SPECIES

4.5.1 *Behaviour of* Rattus rattus

Barnett [19] has studied the behaviour of *R. rattus* in the same
conditions as those in which the social behaviour of *R. norvegicus* was
observed. Most of the major behaviour patterns are the same in the
two species. *R. rattus* is the 'shyer' of the two: it is more easily disturbed by an experimenter and more mercurial in its movements.
Males attack male newcomers of their own species and also *norvegicus* of
either sex; but attack cannot be predicted with as much certainty as in
norvegicus. There is no evidence of differences of social rank in stable
colonies.

Although it is sub-tropical in origin the tendency to huddle in
groups is as marked in this species as in *norvegicus*. All the components
of amicable contact behaviour seen in the one were observed also in the
other, but the movements of *rattus* were more rapid and graceful –
a reflexion, perhaps, of its climbing habit. Fighting is again a male
prerogative. Tooth-chattering is conspicuous in some individuals,
absent in others. The threat posture and mincing movements are the

same, but attack is more often opened without preliminaries, by leaping and biting, than in *norvegicus*; the leaps, too, are longer. Biting is only momentary, as far as is known. 'Boxing' often occurs in the intervals between more violent episodes. As in *norvegicus* fights are always between two individuals only, although a newcomer may be attacked by two or more resident males in succession. A difference from *norvegicus* is that during a violent fight the excitement sometimes 'irradiates' among other members of the colony which run and jump wildly about.

33. Interspecific conflict (an imaginary scene). The artist shows, in this dramatic drawing, a male *Rattus norvegicus* (*right*) seizing and biting a male *R. rattus*. Despite its apparent improbability, events of this sort have been observed and filmed in the laboratory (plate 8). (From Vogt & Specht. Die Saügetiere. 1883.)

4.5.2 *The two species together*

As already stated, *R. rattus* males usually attack *R. norvegicus* adults which enter their territory; similarly, *norvegicus* males attack *rattus* of both sexes. These attacks occur even though a member of one species may perform 'amicable' acts directed towards a resident (figure 23). Certain aspects of recognition may be studied in situations in which the two species mix. First, conflict is not inevitable: we may assume that the odour of members of one species differs distinctively from that of another, but this in itself does not provoke attack. If adult males of both species are introduced all at the same time into a large cage, they may live together in good health indefinitely, huddling together during sleep and feeding peacefully together at other times. As with intraspecific relations in *norvegicus*, it is a territorial situation that leads to fighting: one male must be an established resident, and he must be faced with a stranger. The stranger may be of either sex if not of his own species; this indicates that the distinctive female odour, which evidently inhibits attack, is different in the two species.

The fact that females not of the same species are attacked gives some further information on the nature of the olfactory stimuli involved. Evidently, while the various amicable and aggressive 'signals' are shared by the two species, the signs on which sexual recognition depends differ. This is what would be expected. It would be disadvantageous if the sexual energies of individuals were directed towards members of another species with which no fertile union is possible. But the barrier of odour is not quite as well constructed as it might be. Male *norvegicus*, in their own territory, do not merely attack strange male *rattus*: they sometimes mix aggression with attempts at coitus. This is *not* observed with female *rattus*. The likely explanation is that male *rattus* have an odour which sufficiently resembles that of female *norvegicus* to confuse male *norvegicus*.

4.6 ANIMAL COMMUNICATION

Many of the general principles governing the forms of animal communication have been illustrated in this chapter. The origin of terms such as 'communicate' and 'inform' is to be found in human social activities and obligations; but in scientific contexts today these words usually refer to some aspects of engineering or mathematics: and now some biologists are trying to use the engineers' concepts as a source of a language for describing and explaining behaviour. They have not yet got very far, but it may well be that this attempt represents the ethology of the future.

Communication between animals involves the passage of information; this depends on *signals*. A signal is a small amount of energy or matter which brings about a large change in the distribution of energy or matter in a system. The definition does not require action at a distance, but the signals we are concerned with here do usually act at a distance: the source of the signal and the system influenced are both animals; the signals are *social signals* and they are of course a class of stimuli. They may (i) have the function of encouraging approach and the performance of some 'coöperative' act, as in mating or the care of young; (ii) they may merely prevent attack; or (iii) they may induce withdrawal. Among rats and most other mammals olfactory signals play a major part in (i) and (ii). As for the third function, the clearest and most studied examples are the sights and sounds produced by birds defending their territories;

but we have seen that territorial behaviour is found also among male wild rats, and Hediger [139] has described many other examples from the Mammalia.

Since social communication involves two parties, it must be the product of a dual evolution: the ability to emit signals must have evolved in conjunction with the readiness to respond to them. This, however, requires qualification: colours, shapes, postures, movements and odours which have become signals must often, perhaps always, have originated in advance of their use in communication. As Darwin showed in *The Expression of the Emotions*, an animal's internal state is often reflected in its appearance, or in the sounds or odours that it emits. These by-products of physiology can give information on the animal's subsequent behaviour: the odour of a female rat (determined by endocrine secretions) can indicate whether she will accept a male or kick him away; the raised hair of a male (reflecting the state of the autonomic nervous system) tells at least a human observer something, though it is uncertain whether it influences other rats.

An important question, to be discussed more fully in the next chapter, is the extent to which learning is involved in the ability to respond to signals. In man, responses to social signals are learnt, and vary from one community to another; (so of course is talking, or the emission of signals, except in infancy: only infants have a repertoire of stereotyped signals, such as crying and smiling). In all other animals which have a vocabulary of social signals, the signals are typical for the whole species and are, like the responses to them, highly stereotyped. This suggests at first sight that the neural equipment involved must develop in advance of use and regardless of the particular conditions in which the individual is reared. Yet we know that in many species quite highly stereotyped signals depend on early learning: many birds sing their characteristic song only if they have heard older members of their own species during their formative months. Hinde [151] has suggested that a central nervous organization possessing some flexibility or learning capacity is more readily evolved than one in which there is complete rigidity of both signal and receiving equipment.

Nevertheless, animal communication depends for its efficiency on a high degree of standardization and the use of a small number of simple signals. It might be expected that this would lead to easy understanding by a human observer, but this is not the case. Our comprehension of fellow-members of our own community (and of members of other communities in so far as we do understand them) depends on a very detailed system of signals, many given and received

unconsciously: not only speech but also posture and slight movements
make their contribution. Above all there are the facial muscles of expression which, with the movements of the eyes (made visible by the
'whites' [1]), convey a great deal of information. A most striking
feature of the behaviour of animals such as rats is the absence of signals
of this sort: there are no muscles of expression, and this is one reason
why rat behaviour often seems to us to be both inconsequential and
enigmatic. The same applies to most other animal species. The domestic
dog (*Canis*), which has evolved for millennia as a commensal of man, is
an instructive exception.

Another contrast with our own behaviour is that the signalling of
animals is performed, nearly always, without regard to the ability of
other individuals to receive the signals. Man has some awareness of the
needs, feelings and so forth, of other people: although he makes many
mistakes, he usually adjusts his behaviour accordingly. By contrast, the
sounds and other signals of other species are made, as we say, 'automatically': alarm calls may be uttered when there is no other animal to
hear them.

A further unique feature of human communication lies in a special
quality of our language. We have already seen that it is not species-
characteristic – obvious enough in view of the immense variety of
human tongues. The great dependence on learning is accompanied by
an ability to report past events and to speak about general classes of
objects. This is not found in other species: their signals express alarm
or an intention, such as readiness to attack or to mate; although animals
learn, they make use of the learning only in guiding their own subsequent action and not to communicate information to others.

4.7 CONCLUSIONS ON RAT SOCIETIES

The social behaviour of rats, as of other vertebrates, may be regarded
as a rather complex system of equilibria between the herding tendency
and disruption. Individuals of the same species must live for the first
weeks of their lives in a family group, and must come together to mate.
Ordinarily, the tendency is, even apart from family and reproductive
behaviour, to approach another rat and to associate with it in close contact. If this were the sum of the social behaviour patterns of rats, a
colony would grow, as fast as its reproductive capacity and immigration
allowed, until food shortage or other density-dependent factors prevented further increase; those rats that did survive in these conditions
would remain together, many of them doubtless in a poor state of health.
However, two other powerfully developed kinds of behaviour break

down the adhesiveness of rat colonies. One, fully discussed in § 2, is exploration; the other is the territorial behaviour of male rats.

Social interactions, whether syncretic or disruptive, involve social 'releasing' stimuli and the responses they evoke. For the purposes of this book the term 'social releasing stimulus' means a visual, olfactory, auditory or other quality or group of qualities of an individual which *either evokes or inhibits* a stereotyped behaviour pattern in another member of the same species. The definition permits the inclusion of stimuli which would otherwise have to be called 'social suppressors', or something of that sort; further the definition, by using the non-committal word 'stereotyped', does not anticipate the discussion of 'instinct' in the next chapter.

In many species, conflict within the group is limited by the formation of a social hierarchy: certain behaviour is permitted, other behaviour is not. There is no evidence for the formation of hierarchies in rats; but relationships of dominance and subordination do occur, at least among the males of experimental colonies: the viable types of male are the 'alphas', which are dominant but of which there may be several in a colony, and the 'betas', which have adapted themselves to a subordinate rôle. One alpha is equivalent to another, and one beta to another beta, as far as observation goes. It is possible that this is a primitive type of social organization, and that hierarchical systems have evolved independently on many lines of descent. The widespread occurrence of these relationships of contrasted herd behaviour and hostility serve to emphasize further the importance of an equilibrium between the behaviour involved in herding and that of dispersal. The facts also show that crowding in itself does not lead to conflict. Where there is conflict a rat may go short of food even though there is plenty of food within its home range.

The differences between an alpha and a beta are a result of individual experience. The rôle of learning in the social behaviour of rats is difficult to define, but the attitude of a beta towards an alpha is an obvious example of it. Other minor examples have been quoted above; but, on the whole, the social activities of rats consist of stereotyped behaviour patterns. The notion of a 'rat sociology', comparable to that of man with his infinitely variable behaviour, is as unjustifiable as the belief that rats have virtually no social life.

The stereotyped character of the social behaviour of rats is made more evident when *R. norvegicus* is compared with *R. rattus*. The motor patterns are closely similar; indeed, they are barely distinguishable. The releasing stimuli are the same in both 'amicable' and 'aggressive' behaviour, but there are important differences in those that evoke coitus: males of one species do not attempt coitus with females of the other. Thus a barrier, probably olfactory, between the species exists at exactly the point at which it would be expected on general grounds.

A coherent account can, then, be given of intraspecific relationships among wild rats. But much remains to be learnt. Among the major problems is that of the way in which the various behaviour patterns develop. Just what is the rôle of experience in the emergence of stereo-typed social activities? The importance of this sort of question is further shown in the next chapter.

5

Fixed Patterns and Heredity

Much that we attribute to instinct, because no *prolonged* learning is evident, might . . . be due to learning that needs only a few seconds for its completion. The associations that are formed may be only certain ones to which the nervous system is especially adapted.

D. O. HEBB

5.1 DESCRIPTION

5.1.1 *General*

It is sometimes possible to predict accurately how a rat will behave, without knowing anything detailed of its previous history: one may need to know its sex, whether it is mature and, if a female, her reproductive state; but one need not know whether the rat has previously encountered the situation in which it is being observed. The prediction may be only of the direction of movement in relation to an object in the environment: for example, towards a burrow or another rat. But detailed prediction can sometimes be made of a complex *pattern of movements*; these are often called 'fixed action patterns'. Every species displays certain standardized forms of movement which are characteristic in just the same way as are structures. (Man as an adult is perhaps an exception: but even in *Homo* walking is standardized.) These stereotyped patterns are often referred to as species-characteristic; but, since they may be shared by related species, it might be better to say that they are taxon-specific. Several examples were given in § 4: all the components of fighting and of amicable behaviour are shared by the two species *Rattus norvegicus* and *R. rattus*; so, probably, are those of maternal behaviour.

Among the questions which may be asked about fixed action patterns are: what stimuli evoke them? how are they controlled by internal processes?; how do they develop in the young animal? But before these can be considered further description of some of the movements themselves is needed.

The most easily observed of stereotyped movements are those of respiration, locomotion and eating. These are so obvious that they are often overlooked, although they help to illustrate the relationship between reflexes and more complex patterns. Respiratory movements are often described as 'reflex'; eating involves the reflex secretion of digestive fluids and movements of the alimentary tract; locomotion depends on a system of 'feedback' proprioceptive reflexes from all the moving parts. The acts called reflexes are little influenced by fluctuations in internal state; they are comparatively simple, although they may involve many muscles contracting and relaxing in a delicately co-ordinated way; they are standardized; and they show little improvement with repetition. In these respects they differ from fixed action patterns only in their relative independence of internal state and in their simplicity, and they are indeed not sharply marked off from them. Reflexes are always components of complex fixed patterns, as they are of all behaviour; and the same reflex may appear in many different complex patterns. But we shall see that many fixed patterns have a quality not possessed by typical simple reflexes: the readiness with which complex activities are performed varies with the internal state of the animal, and this may depend on the lapse of time since the act was last carried out. This is obviously true of eating. Simple reflexes are usually set off by an immediate, standard, sensory input (as when salivation is evoked by food in the mouth). Complex stereotyped behaviour patterns become decreasingly selective in the stimuli which evoke them, as time elapses since they were last performed (that is, with increasing 'drive') [145].

The movements of respiration illustrate the fact that, nevertheless, the boundary between complex patterns and reflexes is blurred. Usually, respiration is left to the physiologists, though the respiratory rate may be used as an index of 'emotional state' during studies of behaviour. But Spurway & Haldane have shown how, in a special case, the movements of breathing can be analysed in ethological terms. They studied newts (*Triton*) resting on the bottom of an aquarium and rising at intervals to gulp air. The movement to the top represents appetitive behaviour and the breathing itself is a consummatory act; the change in blood composition, of course, leads to a consummatory state. Spurway & Haldane discuss not only the overt behaviour of breathing but also the internal changes which influence the behaviour. The control of breathing depends on negative feedback: the greater the rate at which oxygen is used and carbon dioxide evolved, the higher the rate of the respiratory movements [300].

The activities mentioned above are necessary for individual survival: respiration and ingestion are directly vital and locomotion is an essential component in, for example, getting food. Such behaviour is sometimes said to be 'homeostatic' in function. Each inspiration is evoked by the slight accumulation of carbonic acid in the blood acting on the respiratory centre in the brain stem: the subsequent expiration removes carbon dioxide and so helps to maintain the carbonic acid content of the blood within narrow limits. This is a standard example of homeostasis, or the maintenance of a steady internal bodily state. The corresponding facts concerning ingestion have been discussed in some detail in § 3.4.

34. Grooming.

5.1.2 *Grooming*

However, other stereotyped activities are not homeostatic in this sense. A conspicuous but little studied example is grooming (figure 34). The following are notes made on the behaviour of a wild rat on waking.

> Eyes open. Rises and stretches. Licks hands; washes face; washes behind ears, continuing to lick hands at intervals. Licks fur of back, flanks, abdomen. Licks hind toes, scratches with them; licks, scratches, licks, scratches . . .; hind toes scratch flank and belly from front to back. Licks genitalia. Licks tail, holding it in hands. Licks hind legs, holding them in hands. Bites fur. [26]

This sequence is typical of both wild and laboratory rats, of both sexes, and of all ages from weaning or before. The behaviour at least *seems* to be performed regardless of need: the animal does not wait until there are parasites or dirt in its hair. The behaviour resembles breathing and eating in that readiness to perform it grows in the absence of any evident special external stimulus, but no experiments have been performed on the factors influencing the build-up of the readiness to groom: perhaps the accumulation of skin secretions is a factor; no doubt the presence of foreign bodies on the skin stimulates grooming. There is scope here for a detailed study.

5.1.3 Nest-building

Grooming may be compared with nest-building. Given access to suitable material a rat will usually make a nest at least of a hollowed-out heap of material into which its body fits; the site of the nest is ordinarily under cover, often in a burrow. The material is carried in the mouth; it may be straw, fragments of clothing or sacking, paper and much else. Paper or cloth are torn up, and cloth and similar material may be fluffed out as well. Kinder found that rats aged twenty days, with no previous experience of nest-making, built a nest of paper within three hours [172]. Rats do other constructional work too. Barnett has observed the filling in of spaces in the walls of wire cages with a mixture of dung and cotton wool, to form a screen (plate 11); this operation is carried out with a rapid, stereotyped patting movement of the forefeet—a procedure which is perhaps important in the construction of burrows in earth [23]. Nest-building, then, with related activities, is a complex system of movements, some of them stereotyped, but as a whole adaptable to a fair range of circumstances and materials. A more detailed account of a comparable sequence has been given by Hinde, in a study of nest-building by canaries (*Serina canaria*) [149].

What are the factors that influence nest-building, apart from the materials available? This question has not been fully studied experimentally. Temperature is certainly important. Kinder gave laboratory rats access to strips of paper, and found that the weight of paper torn off for nest-making increased as the temperature was lowered [172]. The building of a screen, described above, occurs only at the front of a cage and on its roof; the factors involved may be currents of cold air, or light. This needs investigation. The homeostatic significance of nest-building has been more subtly displayed by Richter, by showing that the nest-making of rats improved after hypophysectomy. Removal of the pituitary causes degeneration of the thyroid, and so an impaired ability to increase metabolism at low temperatures. Richter also gave thyroxine to hypophysectomized rats and found that this prevented the effect of the operation on nest-building. Thyroidectomy had an effect similar to that of hypophysectomy [254]. Figure 35 shows a form of Skinner box in which depressing a panel gives the rat a source of heat;

rats in this apparatus learn to adapt their behaviour to their need for warmth.

Nest-building is not always a function of external conditions. Females build a nest shortly before parturition, and this behaviour occurs in a wide range of temperatures. Gelineo & Gelineo, however, find that parturient females, if given a choice, make their nest at a temperature

35. Skinner box in which the reward of pressing the panel is the switching on of a source of heat. Rats adjust their rate of response to their need. (After Weiss [329].)

below that of thermal neutrality. They believe that this favours the growth of the young [110]. Wiesner & Sheard showed that the nest of a pregnant rat is usually maintained during lactation, though to a diminishing extent during its later stages. Here the behaviour is influenced by the reproductive state of the female, evidently through the action of hormones on the central nervous system [330].

Nest-building, then, is homeostatic in that it contributes to the maintenance of a constant body temperature, and is altered appropriately with changes in the temperature of the surroundings. At the same time there is, in a female, an 'autonomous' process which brings about nest-building. The nest built by a pregnant female may contribute nothing to her own thermoregulation, and so is not homeostatic in this sense. These examples illustrate the fact that fixed action patterns are diverse both in their functions and in their causes. This is further brought out in the list of behaviour patterns given on page 106.

TABLE OF STEREOTYPED MOVEMENTS

1. Respiration	Not a complex sequence; not very variable; performance depends on build-up of an intetnal state (§ 5.1.1)
2. Locomotion	Several types, according to speed and terrain; includes climbing, swimming; necessary component of, or preliminary to, other stereotyped acts
3. Ingestion	Includes (apart from locomotion) handling, sniffing, licking, gnawing, salivation, swallowing; sucking in infant; depends on build-up of internal state (§ 3.2); influenced by early history
4. Gnawing	May be performed independently of ingestion or any evident need
5. Hoarding	Includes picking up, carrying, dropping; causation complex (§ 3.2.4); influenced by early history
6. Grooming	Complex, highly stereotyped in absence of specific irritation; performed 'spontaneously' (§ 5.1.2)
7. Crawling under; grooming others	Amicable behaviour, including several independent components; evoked by external conditions (§ 4.2.1); influenced by previous experience of fighting
8. 'Fighting'	Several independent components (including 'phonation'); depends on external stimulation (§ 4.3.3); influenced by previous experience of fighting
9. Coitus	Complex sequence; internal build-up (§ 5.2)
10. Parturition	Complex sequence (§ 4.2.2)
11. Nursing	(§ 4.2.2)
12. Retrieving	(§ 5.2.2)
13. Nest-building	'Spontaneous' if materials available; but influenced by conditions; also build-up in pregnancy, followed by gradual decline after parturition (§ 5.1.3)

5.1.4 *Sensory abilities*

Since fixed action patterns are often performed in response to a highly specific pattern or quality of stimulation, there must be correspondingly fixed sensory arrangements: the stimulus must in some way match a special arrangement of neurons in the sensory part of the central nervous system. In § 4.3.3.2 examples were given of social signals which seem to have fixed effects regardless of experience. There is indeed sometimes a sensory organization without a corresponding special motor pattern. Admittedly, since we usually have no neurological data, we know of the sensory side only through observation of behaviour; but the behaviour, instead of consisting of a specific pattern like coitus or 'fighting', may be only the adoption of a particular direction of movement or even the cessation of activity. For instance, a rat may become quiescent on attaining a particular skin temperature.

A possible (but not proved) example of a stimulus which produces a non-specific, non-learned response in rats is the odour of a cat: Griffith, and later Curti, have both suggested that rats 'freeze' on their first encounter with a cat [116, 78]. Better examples of non-specific responses are given by the exploratory and avoidance types of behaviour described in § 3; for these, however, the stimuli are themselves non-specific: exploration and avoidance (as we saw) are evoked by change, and not by any fixed pattern or mode of stimulation. As far as is known, the propensity to respond to change is developed regardless of special experience.

Another kind of sensory ability, still non-specific, is the perception of distance. A terrestrial animal, if it is above a certain size, must be able to avoid falls as soon as it becomes active; if it depends substantially on vision for getting about, it should be able to judge distances, especially vertical ones. In 1934 Lashley & Russell studied this question with an apparatus in which rats were required to jump from a stand to a platform of which the distance could be varied (figure 36). Rats which had been reared in the dark were able to jump as accurately as control rats [190]. However, the dark-reared rats needed a little experience of the light before they would jump, and so the experiments were not quite conclusive.

This difficulty has been overcome in the work of Gibson and others: they have shown that children (*Homo*), goats (*Capra*) and kittens (*Felis*) probably possess an innate perception of vertical distances. In these experiments a 'visual cliff' was used: a narrow platform has plate glass flooring on each side; on one side a patterned floor is just below the glass, while on the other side the floor is a foot or more below, giving the appearance of a 'cliff'. Hooded rats (unlike the other species mentioned) showed little preference for the two sides, provided they could feel the glass with their vibrissae; but if the central platform were raised several inches, so that the glass could not be felt, the rats nearly always walked on to the shallow side. If there was no visible pattern on either side, no preference was shown.

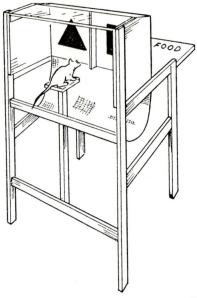

36. Apparatus for experiments on visual discriminations. There are two doors, marked with different patterns. One is unlocked and gives access to food when the rat jumps at it. The other is locked: a rat that tries to get through falls back into the net below. (After Lashley [186].)

The ability to respond to distance has been called above a 'non-specific' sensory ability, but this does not absolve us from trying to find out just what feature of the situation is responded to. There are at least two possibilities. The further away a pattern is, (i) the more elements of the pattern are seen at once, (ii) the more slowly the elements move across the field of vision when the animal moves its head. In rats it is evidently the second, that is, motion parallax, which is independent of learning: even rats reared in darkness could make use of this feature of the situation as soon as they were put, in the light, on the experimental platform [113].

The examples in this and the preceding sections show then that rats, like other animals, possess a repertoire of stereotyped behaviour patterns and of sensory arrangements which (*a*) have obvious survival value and (*b*) seem to be as fixed in development as 'structural' features such as the shape of the ear or tail (both used in classification) or 'chemical' effects such as hair colour. There remains, however, much to be said about the agencies, external and internal, which control the stereotyped activities; we turn now, therefore, to the most studied group of all, those involved in reproduction.

5.2 REPRODUCTIVE BEHAVIOUR

5.2.1 *Mating*

5.2.1.1 *Description.* The extensive literature on reproductive behaviour in rats has been surveyed by Beach [37], Larsson [184], Munn [233] and Thorpe [312]. When references are not given in the following account, they will be found in these reviews.

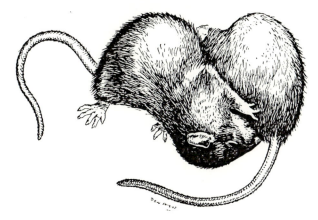

37. Mating. A male (*left*) investigates the external genitals of a female shortly to come into oestrus.

When an adult male, recently deprived of female company, encounters a female, he approaches and sniffs her; he may sniff and lick her genitalia, adopting a characteristic posture with head pointed down (figure 37), in order to do so, or he may omit this formality and try to mount her. If the female is not in oestrus she does not allow intromission: on the contrary, typically she kicks the male off (plate 9); or she may merely walk away. If the female is in oestrus she may herself take the initiative in approaching the male and nuzzling him (figure 38). Whether she does this or not, after the first contact she runs a short distance away and pauses; the male follows and mounts her; as he does so the female adopts a position which permits intromission, with the coccygeal region raised and the tail to one side (figure 39). If the complete sequence is achieved the male inserts his penis, ejaculates and immediately leaps backwards, all in less than three seconds. Ejaculation, however, at least in laboratory rats, occurs in only about twenty per cent of copulations. In any case, the male now sits back on his haunches and licks his penis. In a few seconds the male resumes, and the whole sequence is repeated many times, at intervals of a few seconds, over a period of an hour or more.

38.   Mating. A female in oestrus (*right*) approaches a male.

This behaviour is characteristic of the whole species, *R. norvegicus*: it has been observed in various laboratory strains and also in wild rats. It differs substantially, however, from the mating of some other rodents: for instance, in the guinea-pig (*Cavia*) the male may rest for more than an hour after a single ejaculation [Young, 342, p. 97]. Whatever the species, the whole sequence is highly stereotyped: in each sex it is an obvious example of a fixed action pattern. We may now ask: (i) what are the signals to which rats respond during mating? (ii) what internal processes are involved? (iii) what factors influence the development of the ability to mate?

5.2.1.2 *Signals.* In the current accounts of stereotyped behaviour, based mainly on studies of insects, fishes and birds, stereotyped acts are often said to be evoked by a single, standard stimulus-pattern or social releaser. Some features of such stimuli have been discussed in § 4. Usually, the stereotyped act is performed only if the specific stimulus is presented, but there are two important exceptions to this rule. First, if the act has not been performed for a long time, it may occur in the absence of the usual stimulus: it may be directed towards a substitute object, as in homosexual behaviour, or it may seem to occur in the entire absence of an evocative stimulus (vacuum activity). Second, 'supernormal' stimuli may be devised experimentally; these are more effective than the usual stimuli in evoking the response: an example is a model egg, much larger than a normal egg, which induces a bird to sit on it in preference to one of its own. These phenomena have been reviewed by Tinbergen [314, 315].

39. Coitus. The raised coccyx and deflected tail of the female make intromission possible. The pressure of the male's forelimbs on the flank is one of the stimuli needed for the female's response.

In mammals, social stimuli act in a similar way, but it seems that as a rule more sensory modalities are involved and that often no one modality is essential. In § 4.3.3 the stimuli which probably evoke attack were described: evidently a number of qualities, visual and olfactory, may act additively to produce the response. This is certainly the case with the mating behaviour of male rats. Study of overt behaviour suggests that olfactory stimuli are of primary importance in all kinds of social behaviour in rats (as in other mammals); and Le Magnen has shown that male white rats can distinguish the odour of a female in oestrus from that of other rats [205]. Yet Beach, following earlier work, showed that mating will take place when olfaction has been made impossible by destruction of the olfactory bulbs of the forebrain: the probability that an anosmic male will copulate is lower than it is for a normal male, but the olfactory stimuli presented by a receptive female are certainly not a *necessary* condition of sexual behaviour in a male.

To what stimuli, then, is an anosmic male responding? Two further sensory modalities were studied by Beach: (i) vision and (ii) the cutaneous sensitivity of the snout and legs. In a small number of experiments two of the three senses were destroyed: for instance, olfaction and vision. If a rat treated in this way had had no previous experience of coitus, he failed to respond to a female; but if he was experienced he still copulated, evidently responding to stimuli acting on the one remaining sense; there was, however, substantial individual variation in the effects of the operations. Males in which all three senses were destroyed did not copulate. Not surprisingly, few of these ruthless experiments have been done; but there have been enough to show that the pattern to which the male responds involves at least the three senses mentioned; and that even some inexperienced males are able to do without one of them. The effects on behaviour of the three kinds of stimulation are additive. The fact that experience influences the behaviour of treated males should be noted.

5.2.1.3 *Internal processes: nervous system.* Whatever the precise form of the stimuli that evoke stereotyped acts, we must assume that there is a standardized arrangement of the central nervous system corresponding to the behaviour; but, since so little is known of the way in which the brain works, it is difficult to find a sensible form of words to express what may reasonably be said about it.

In animals with simple nervous systems behaviour is usually an immediate reflexion of the stimulus situation. (By 'the stimulus situation' is meant, not the total situation in which the animal finds itself, but only certain features of it: the sense organs and nervous system of each species are equipped to respond only to certain types and patterns of change in the environment. Rats do not respond to the colours of objects, nor to sounds of low pitch and intensity, though men, birds, some fishes and many invertebrates do so.) The direct sensory control of behaviour is reduced with greater volumes of neural tissue: the amount and complexity of the mass of connecting or internuncial neurons influences the extent to which behaviour is influenced by past experience rather than the stimulus of the moment.

The division of the cells of the nervous system into sensory, connecting and motor neurons is familiar from texts on anatomy and physiology, but it does not coincide with a division of neural function based on observing behaviour. Connecting neurons, in particular, evidently have at least two kinds of function. First, they may act merely as part of

9. Rejection of a male. A female not in oestrus kicks off a male trying to mount her. (1/1250 sec.)

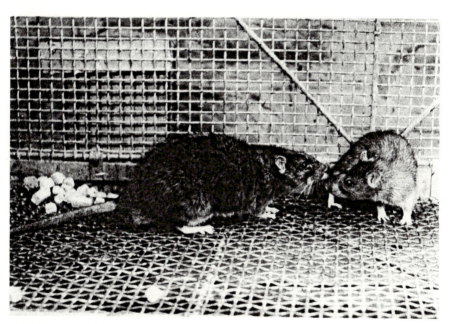

10. Mating. A large, dominant male with hair slightly raised approaches a strange female.

11. Nest-building. A wild female with young has made a screen from a mixture of cotton wool, dung and peat moss stuck to the wire wall of the cage.

a system, fixed in advance of use (during normal development) which ensures the appearance of stereotyped behaviour. The best studied of such neurons in mammals are those of the spinal cord; most spinal reflexes (which are highly stereotyped) involve at least two sets of synapses: one between sensory and connecting neurons, the other between connecting and motor. Secondly, other connecting neurons, especially (in a mammal) those of the forebrain, introduce a much greater complexity into behaviour and reduce its dependence on the immediate situation. But it must not be assumed that there are two *kinds* of neuron, one with fixed connexions (giving rise to stereotyped behaviour) and the other with variable connexions (giving rise to variable behaviour). Too little is known about the nervous system to justify any such assumption. Indeed, even in the spinal cord synaptic changes take place with use [90]. Further, as we shall see more fully below, complex stereotyped behaviour, however fixed the pattern seems to be, often depends on, or is influenced by, individual experience. A wide range of fixed patterns can be observed in animal behaviour, from those which are remarkably stable in a great variety of environments to those which are highly labile [151].

Where behaviour is highly standardized, as in the mating of rats, the neuro-sensory arrangement involved has sometimes been named an 'innate releasing mechanism' or IRM; but the term 'innate' represents an assumption which is often unjustified. As a rule, nothing at all is known about the specific mode of operation of the neurons involved: we cannot point to particular groups of cells in a rat's brain and state that the excitation of these cells according to a particular pattern will bring about, say, the movements of coitus.

5.2.1.4 *Internal processes: endocrine.* Nor can we identify particular chemical processes going on in neurons, and say that given behaviour patterns depend on them. We take it for granted that there are chemical changes going on; and a good deal is known of those which constitute the nerve impulse in peripheral axons; but little can be said about what goes on within nerve cell bodies, or of the chemical differences between one cell type and another.

There are, however, substances which, we know, must reach a specific level in the blood before certain activities are performed. These substances evidently influence the activity of neurons in the brain. The most familiar example (already mentioned above) is the action of carbon dioxide on the respiratory centre in the medulla: the rate of breathing movements is largely determined by this chemical agency. This is an example at the reflex level.

More complex stereotyped acts depend on hormones. In mammals generally sex-specific behaviour usually occurs only if the gonads are secreting certain hormones; and gonadal activity depends in turn on pituitary secretions. This holds not only for mating behaviour but also for care of the young and for fighting.

Castration of a male reduces or abolishes sexual behaviour. This holds whether the operation is performed when the rat is a day old or when it is an adult; but in the latter case the effect may appear only gradually: at first, behaviour may be unaffected. Injection of substances chemically similar to androgenic hormones produced by the testes restores normal sexual behaviour (figure 40). The substance usually

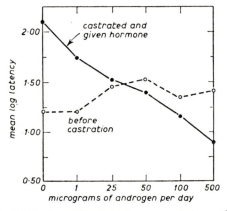

40. Hormones and behaviour. The behaviour recorded was the length of time (latency) before first coitus after introduction of a female to a male. The broken line indicates the latency of control males; the continuous line refers to castrated males given different doses of male hormone. Latency declined with increasing doses. (After Beach & Holz-Tucker [40].)

employed is a salt (the propionate) of testosterone; testosterone is one of the steroid hormones which can be extracted from testes. The same substance, given to males aged twenty-nine days, induces precocious sexual behaviour; normally, sexual behaviour does not appear before the age of thirty-five days.

In the female there is a parallel effect of hormones secreted by the ovaries (figures 41, 42). If a young female is spayed, normal sexual behaviour does not appear; in a mature female spaying is followed by permanent anoestrus and invariable rejection of the male. Responsiveness is restored if suitable amounts of both oestrogen and progesterone are injected.

The extensiveness and unequivocal results of research on these hormonal effects has made them loom large in accounts of the physiology of behaviour. Certainly, they are interesting and important from many points of view. Endocrine effects are, however, peripheral to the most difficult problems – the problems of the nervous control of behaviour. One question is how hormone action is related to neural function.

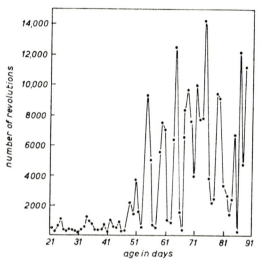

41. Hormones and behaviour. The four-day cycle of activity, as recorded in an exercise wheel, corresponding to the oestrous cycle, begins at puberty. It is irregular at first. (From Wang [326a].)

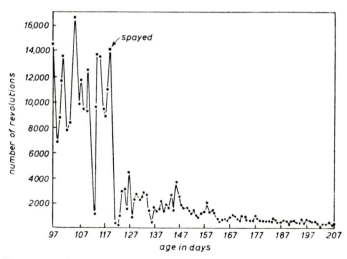

42. Hormones and behaviour. The four-day activity cycle was ended by removal of both ovaries. (From Richter [255].)

The obvious answer is that hormones activate already existing neural arrangements. The facts summarized above suggest that the central nervous organization for mating behaviour is present, but is ineffective until it is set going (how, we do not know) by steroid hormones. There is some evidence that these hormones increase general excitability: in other words, they lower all the thresholds which bar the entrance of sensory inputs to the executive departments of the central nervous system. For example, as already mentioned (§ 2.2.2), a female rat in oestrus is much more active in a running wheel than she is at other stages in the cycle. Again, testosterone increases aggressiveness in the male as well as sexual readiness.

However, all this is very vague: what do we mean by 'activation'? Unfortunately, we have at present no information to help us on this point. A second question concerns the stage between stimulus and response which is affected by the activating hormones. The preceding paragraph implied that the effect might be on the sensory side of the nervous system. Le Magnen has shown, not only that male rats can distinguish the odour of females in oestrus but also that this ability disappears after castration and can be restored by administering testosterone [205]. Here is an example, then, of a specific sensory function controlled by a hormone. It is possible that the other qualities of a female (visual and tactile) which evoke mating behaviour in a male are distinguished only if the male hormone is present. But we cannot yet say more than this.

A third question, discussed by Lehrmann [194] in a valuable review, is whether the effects of sex hormones are always central. Could the primary action be on organs outside the central nervous system? Certainly, the development of the secondary organs of sex in mammals depends on the gonadal hormones; but sexual behaviour can take place without most of them: male rats without seminal vesicles, coagulating glands or prostate mate normally; even castrated males will do the same if they are given testosterone.

However, this is not the whole story. Among the secondary sex characters in the rat are papillae in the skin of the penis, and these too develop only if there is enough androgenic hormone in the blood. Beach & Levinson have shown that the extent to which these papillae are developed influences the amount of mating behaviour performed by males which have been castrated and then given male hormone. Hence the intensity of sexual behaviour depends in part on this peripheral effect; the papillae themselves are, of course, sense organs, and behaviour is influenced by them via the nervous system. This does not at all preclude a direct action of the hormone on the brain as well; and the authors quoted suggest that the action on the papillae is responsible for only part of the effect of the hormone on behaviour [43]. Another instance of peripheral action is found in the female: ovariectomy not only prevents response to a male's advances; it also leaves the female unattractive to males, evidently through the failure of an odorous secretion. Lehrmann himself, in an elegant study, has described in detail peripheral effects of prolactin in the ring dove, *Streptopelia risoria* [193].

A fourth question concerns the relationship between hormone action and the effects of previous experience. So far, the discussion has been in terms of a stereotyped activity being, so to speak, switched on and off; and this might be taken to imply that the neural organization responsible is itself completely fixed. But the brain is an organ of which one major function is to change performance according to individual experience. This point is particularly well illustrated by the sexual behaviour of cats. Rosenblatt & Aronson castrated male cats and showed that the effect depends in part on the animal's previous experience: a cat that has not copulated before is likely to be sexually unresponsive after operation, but in an experienced animal the mating pattern will probably persist. 'Once organised through experience, this pattern becomes partly independent of the hormone' [262].

Learning evidently alters the effect of androgenic hormone on behaviour. The hormone itself also influences the propensity for learning: the inexperienced animal first achieves mating and coitus only if the hormone is present; thus the hormone not only puts a stereotyped pattern into operation, but also, in doing so, creates the conditions in which a kind of learning takes place.

Finally, the influence of the brain on the endocrine glands must be considered. The causal chain is far from being one-way: as must be expected on general grounds, the hormones affect the brain but the brain also influences the secretion of hormones. The central example is the control of pituitary secretion by the brain. In female rats the pituitary cycle that controls the ovaries is itself under the influence of the hypothalamus: hypothalamic lesions alter the output of both gonadotrophic hormones. This is the physiological basis, albeit imperfectly understood, of the autonomous behaviour cycle which accompanies the sequence of changes in the ovaries and uterus. The oestrous cycle in a rat is typical of that of many small mammals: it is short (four to five days in laboratory rats) and largely independent of season, though it may be affected by low temperatures. The

behavioural changes include not only the responsiveness to the
male which accompanies oestrus, but also much greater spontaneous
activity during oestrus (figure 41). The nature of the 'internal clock'
which maintains this cycle without external prompting is quite
unknown.

To sum up, the internal processes concerned in mating behaviour
can at present, for the most part, be only dimly guessed at: what little is
known, is fragmentary. We take it for granted that there is a discover-
able neural organization behind the observed behaviour; and we know
that the behaviour depends also on chemical agents, the hormones of
the pituitary and gonads. In § 9 some additional facts are given bearing
on the rôle of the brain in controlling this behaviour.

5.2.1.5 *The ontogeny of mating behaviour.* When it is said that certain
behaviour is stereotyped, or that (say) the mating of a male rat is an
example of a fixed action pattern, this does not tell us how the behaviour
develops in the individual animal. We have seen that in male laboratory
rats with access to a receptive female mating appears at about thirty-
five days. This depends on maturation of the pituitary-gonad system
and the consequent secretion of substantial amounts of male hormone
by the testes. The behaviour can be induced precociously by injecting
testosterone. A question that may be asked is: if a male survives to
five weeks, will he make the typical responses to the female regardless of
the exact conditions in which he has been reared?

This question is in fact not answerable. We cannot subject young rats
to all the possible conditions which permit survival, and see whether
any of them interfere with the development of mating ability. All we
can do is observe the effects of selected departures from the usual conditions of upbringing. When this is done, the rat's mating behaviour is
found to be an example of a notably stable fixed action pattern, in both
sexes. The most commonly used special environment is one in which the
animal has no experience of its own species after weaning: usually, the
only living beings it encounters between weaning and adulthood are the
experimenters. Rats so treated respond appropriately when, as adults,
they meet adults of the opposite sex. Inexperienced males, given a
choice between cavies, male rats, female rats not in oestrus and female
rats in oestrus, rarely make mistakes.

A more complex type of experiment has been carried out by Kagan
& Beach, in order to find out whether experience of other rats in early
life could at least exert some quantitative influence on mating behaviour
in the adult. They reared over one hundred male albino rats alone in a
cage from weaning; about half of these were put for ten minutes,

once a week, with a sexually receptive female; fifteen were put instead with another male; the remainder had no encounters with other rats at all. It was found that the frequency of ejaculation, when this was tested later, was *lower* in both the groups of which the members had had regular experience of other rats. These two experimental groups had developed habits of 'playful wrestling', such as occurs usually among members of a litter, and this was held to reduce the ejaculation rate [164]. However, Larsson has suggested that the rats were in a low state of sexual excitability, owing to transfer to another cage; in his experiments, carried out in a slightly different way, he rarely observed behaviour which interfered with coitus [184]. Whatever their interpretation, such observations at least remind us not to regard the development of fixed action patterns as simple.

These facts, with the evidence from the effects of castration (§ 5.2.1.4 above), do, however, give an impression that mating in rats is a rather mechanical affair. We know nothing of the effects of the nestling's experience during its first few days of life; but, at least after weaning, the rat can do without encounters with its own species and still remain sexually effective. In this it resembles many non-mammalian species, but among the mammals that have been studied it is atypical. Valenstein and his colleagues find that cavies (*Cavia*) are adversely influenced by early isolation [322]. Work on species with more labile sexual behaviour than that of rats shows that, while previous experience is not needed for the *arousal* of sexual responses, their effective *organization* depends to a substantial extent on learning. Young has pointed out that not only in the cavy (*Cavia*) but also in the dog (*Canis*), cat (*Felis*) and chimpanzee (*Pan*) early experience plays an important part in the development of the stereotyped behaviour of mating [342], much as it does in man (*Homo*). It seems likely that in this respect the rat represents an early, that is, primitive stage in the evolution of the mammals: evidently, with increasing complexity of the nervous system and of behaviour, found in the Carnivora and still more in the Primates, the development of fixed action patterns became more labile – and so more accessible to experimental upset.

5.2.2 *Maternal behaviour*

The behaviour of a female towards her young can be analysed in the same way as that of a male towards a receptive female [194]. A short description of this behaviour has already been given in § 4.2.2. Recent studies by Beach & Jaynes illustrate some of the main principles. In one series of experiments the signals or sensory cues which evoke the retrieving of young were studied. The behaviour of nursing females in a standard situation was watched: in some experiments the females' senses had been damaged, while in others the properties of the young had been altered, for instance by smearing them with an odorous substance. The properties of the young which helped their mothers to retrieve them included visual, chemical and tactile components; their temperature also had some effect. (Other evidence, such as that of Wiesner & Sheard mentioned above, had already suggested that the squeaking of her young influences the female). None of the 'sensory modalities' was indispensable [42]. Thus, as with mating behaviour in the male, so with retrieving of young we find multisensory control of stereotyped behaviour.

Another feature of the young which influences retrieving is size. The work of Wiesner & Sheard not only demonstrated this effect but also illustrated substantial variation between individuals in the performance even of a stereotyped act. Retrieving was studied immediately postpartum, and some females responded even to young aged as much as six weeks. Usually, the tendency to retrieve declines markedly with the growth of the young. If an old litter is removed and replaced by a younger one the intensity of retrieving is raised. In some females repeating this procedure prolonged the period during which retrieving was performed to over four hundred days [330].

Something can also be said about the internal processes that influence maternal behaviour. First, there is the question of peripheral effects. Maternal behaviour is unaffected by destruction of the mammary glands: hence it is not controlled by the engorgement of the glands with milk. However, as with sexual behaviour, there is a form of 'peripheral' control, namely, that exerted by hormones. Care of young in the rat depends on the action of a hormone on the central nervous system, though not, in this case, an ovarian hormone: ovariectomy does not affect maternal behaviour. The endocrine gland concerned is the pituitary, and the hormone is prolactin. If prolactin is injected into virgin females which have hitherto displayed no maternal behaviour, they develop the retrieving response and other behaviour patterns typical of a female with young; the intensity of the behaviour is proportional, up to a point, to the amount of prolactin injected. Prolactin has also been injected into males, and they too develop maternal action patterns, though not to the same extent as females.

These observations show that the neural arrangements which bring about the motor patterns of maternal behaviour are present in both sexes. Ordinarily, prolactin is secreted only by the female's pituitary, and so the maternal potentialities of males are not realized. Occasionally, however, a male spontaneously develops this kind of behaviour: such anomalies are well known in laboratory rats and are occasionally seen also in wild ones [23]. (Compare plate 13.)

As would be expected, hypophysectomy is usually followed by failure to behave maternally. Nevertheless, there are exceptions to this rule, and these are as yet unexplained. Probably, several hormones and several endocrine organs are involved, and there is a good deal of individual variation in their functioning.

The preceding account of maternal behaviour is very incomplete. It says nothing of nest-building or giving milk (though these have been mentioned earlier). A more serious defect is that it gives an impression of fixity which, even more than with mating behaviour, is unmerited. A much quoted report by Riess suggests that the nest-building of a pregnant female depends on a specific kind of experience in early life. Riess reared rats from birth in conditions in which they had no opportunity to carry any objects at all: they had no food pellets, no nesting material, not even faeces since these fell through the floor mesh of the cage. These rats, otherwise apparently normal, failed to build nests when they became pregnant, although they were given access to suitable material [256]. These very interesting observations have been only briefly described, and it would be well if they were repeated and extended. A criticism of them has been made by Eibl-Eibesfeldt: the tests of nest-building propensity were made over short periods, in an unfamiliar cage, and these are conditions in which exploratory behaviour is liable to be given priority over everything else [91].

Another, analogous study, by Birch, may be less open to criticism, but still seems to have been only incompletely reported. Female rats were raised from infancy with rubber collars ('Elizabethan ruffs') round their necks; these prevented the usual licking or sniffing at the genitalia and other parts of the body. The collars were not removed until the female was just about to give birth to her first litter. Most females so treated failed to care for their young but ate them instead. The suggestion is that early experience of self-licking and sniffing is needed if the female is to respond properly to the stimuli offered by her new-born young [50]. Once again, then, we see that, even though a behaviour pattern may be highly stereotyped in the adult, its ontogeny may depend on special circumstances of the animal's development – circumstances which can be exposed only by experiment.

5.3 HEREDITY AND BEHAVIOUR

5.3.1 *Genetical variation*

Although the behaviour we are concerned with in this chapter is to some extent modifiable, and often has a complex development, it is still true that it is highly uniform within a given species. When some feature of an organism is general to a whole species it is sometimes said to be 'determined by heredity', or, simply, 'inherited'; a more recent kind of expression is 'carried in the gene code'. We must now examine with some care just what can properly be said about the heredity of behavioural traits.

A first general principle is obvious enough when it is made explicit: every kind of character or trait is influenced by the genes. In practice, for convenience, geneticists have often preferred to study clearly visible and sharply distinct differences, such as those in coat colour or abnormalities such as the presence of extra digits. Only rarely can anything be said of the biochemical processes by which the genes influence the development of such characters; but it is often possible to infer with complete confidence that, say, one rat has a white coat and another a brown one because of a difference at a single chromosome locus.

Other kinds of variation with an obvious genetical basis depend on differences at many loci. It is certain, for instance, that wild rats differ genetically from tame laboratory rats; and it is probable that the docility of tame rats is due in part to departures from the 'wild-type' at several loci. This may seem to be too obvious to be worth saying: what else could the behavioural differences be due to? The answer, in fact, is that they might be due to differences in environment, that is, in the external influences that act on the individual from fertilization onwards. Similarities between members of a population may be due not only to the possession of genes in common, but also to sharing a common environment (including the uterine environment). It is a familiar fact, though it has not been much studied, that laboratory rats of a thoroughly tame stock behave in a very intractable way if they are handled after they have been allowed to 'run wild' for a time. 'Wildness' may be defined in terms of flight from the experimenter, while 'savageness' signifies struggling, biting and squealing when handled. In this sense wildness and savageness, or their opposites, tameness and docility, can evidently be influenced by the environment. Similarly, wild rats can be tamed by regular handling from an early age [26].

Strictly, we can be confident that wildness is influenced by differences of genotype only if animals which differ in wildness have been reared in identical conditions. Stone and his colleagues reared wild and laboratory rats, and the offspring of crosses between these two, in identical conditions from the time of weaning: the unmixed wild rats were, as expected, the wildest, the laboratory rats were tame and the hybrids intermediate [305]. For the purpose of distinguishing the part played by the genotype, a more satisfactory kind of experiment would be to exchange some of the young of pairs of females, one wild and one tame, at birth; some 'wild' young would then be reared by a tame foster-mother, and some tame young by a wild one. The young that had not been transferred would be the controls. However, even this is not the perfect design: to make the environment uniform from fertilization it would be necessary to exchange some of the newly fertilized eggs between the two kinds of female, or to rear all the eggs in standard conditions *in vitro*. Both procedures present excessive technical difficulties. It is perhaps unlikely that conditions in the uterus differ between varieties in such a way as to influence wildness differentially. We may in fact be justified in accepting that wild and laboratory rats differ genetically in regard to behavioural traits, as they do in the causes of hair colour and much else; but we must make the proviso that there may be much still to be learnt about the environmental influences on wildness, especially in early life.

That the notion of prenatal influences on behaviour is not merely fanciful is shown by a number of recent studies, reviewed by Thompson [309a]. The effects of unfavourable conditions ('stressors') acting on pregnant females have been especially studied. It has been found that these can influence not only post-natal growth but also behaviour, including 'emotionality' and learning ability. The changes are not necessarily unfavourable: they evidently depend in a complex way on the intensity of the 'stress' (compare § 8.4.2). There is also evidence that such effects can occur in man.

The tameness of laboratory rats is evidently the result of selection by research workers or other breeders, but the early history of this process has not been recorded. There have, however, been formal experiments on the effects of selection on behaviour. Tryon took a population of 142 heterogeneous laboratory rats and selected from them, over eight generations, for the two opposite characters of 'maze-brightness' and 'maze-dullness'; the tests were carried out in a standard maze, in the dark. The result is illustrated in figure 43: he produced two strains, one quick to learn the maze and one slow. This result illustrates the genetical influence on behaviour.

Further observation of Tryon's strains, by Searle, has illustrated a less obvious, but important principle: a strain, selected to display 'intelligence' in one kind of maze, does not necessarily do so in another. He found for instance that the 'bright' rats were more influenced by food reward and were less distractible than the dull ones. In general, maze-learning ability, like 'intelligence' in man, has a number of rather diverse components [280]. Krechevsky had already shown that the 'dull' rats used visual cues more than the 'bright' ones [180]. Clearly, in such a case, the rats' performance will depend on the kind of stimuli involved in the learning. Here again, then, as with wild and tame rats, the environment must be held constant if genetical differences are to be established.

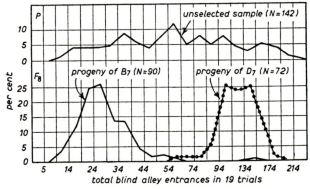

43. Selection for 'intelligence'. The top graph shows the distribution of 'intelligence', measured by the number of mistakes made in a maze, in 142 rats in a mixed stock. From this stock selective breeding was carried out for maze-brightness (B) on one line and for maze-dullness (D) on the other. The result is shown in the two lower distribution curves. (After Tryon [320].)

Most research on behaviour has ignored or, as far as possible, eliminated genetical variation. Yet the work described above, and other studies such as those described in § 4.4, and a few on mice (*Mus*) and dogs (*Canis*), show that there is plenty of scope for investigation [265]. Further, the work of Lát has brought out the potential value of deliberately using genetically heterogeneous stocks. He assessed excitability in rats by recording the speed of acquisition of conditional reflexes and the intensity of exploratory behaviour; and he related these parameters to metabolism. The more excitable animals ate more. They also ate more carbohydrate, when given a choice, while less excitable animals took more protein and fat [191]. This kind of research, combining observation of overt behaviour with physiology and applied to genetically different animals, may well be important in the future. It would be desirable, for example, to relate single-gene differences both to bio-chemical characters and to behavioural traits.

5.3.2 *The development of behaviour*

A further general principle concerns the development of behaviour. Examples of the ontogeny of fixed action patterns have been given above (§§ 5.2.1.5, 5.2.2). It is sometimes said that such fixed patterns are 'innate'. (Often, the term 'instinctive' is, or seems to be, used as a synonym for 'innate'.) The word 'innate', equivalent to 'inborn', *suggests* (however it is formally defined) the meaning of fixed from birth or even from fertilization. We may now ask to what extent one may suit-ably assign the term 'innate', with this implication attached to it, to fixed action patterns.

Let us take again the example of the urinary reflex. In young rats, when the bladder reaches a certain fullness, a sphincter muscle relaxes and the urine is discharged. We saw in § 4.2.2 that this highly stereo-typed reflex, essential for life, does not appear in the new-born rat until the skin is stimulated in the perineal region. One way of putting this is to say that the neural organization on which the response depends completes its development only if the environment provides this particular stimulus. Here is a simple and striking instance of the way in which the development of behaviour, as of all other features of an animal, depends on a continuous interaction of the organism with its environment. This is the principle of epigenesis: development is not a mere unfolding of an organization already laid down; it is a result of a complex interplay between the material with which the organism is endowed at fertilization and the many influences which act on it from then onwards. All development is influenced by both 'nature' and 'nurture'.

This interaction with the environment has already been exemplified in § 3, in the discussion of exploratory behaviour and the avoidance of strange objects. In both the form of the response depends, in a quite specific way, on previous experience: investigation of, or flight from, novelty requires that there must be something more familiar with which to compare the new thing; yet both occur 'spontaneously', without any gradual learning by practice. Such behaviour does not fit into any simple classification of actions into two kinds, 'innate' and 'learned'.

The word 'innate' is most conveniently used, if it is used at all in ethology, as a synonym for 'genetically determined'. This phrase applies, not to 'characters' as such, but to *differences* between indivi-duals. Every 'character' is a product of a complex development, in which genetical and environmental influences interact. But it is certainly true that some differences between individuals are directly related to differences in their genes: in the field of behaviour this applies to the example already given, of 'maze-brightness' and 'maze-dullness'.

How can fixed action patterns be fitted into this framework? To call them 'innate' begs the question of just how they develop: one wants to know, not only what standardized form they have when complete, but by what processes they achieve that form. Sometimes it seems (though this cannot be proved) that fixed patterns will appear if the animal develops at all: it is as if no non-lethal alteration of the environment can stop them. Such patterns may be said to be exceedingly *stable*; others, of which the development can be easily upset by environmental means, may be said to be *labile*. But, even where the development of a pattern is very stable, there must be an interaction between the animal and its environment to bring that development about.

5.3.3 'Lamarckism' and behaviour

In the preceding discussion the conventional 'Darwinian' assumption has been made on the relationship between the experience of the individual and the genetical properties of its offspring. It has been taken for granted that the effects of the environment on the phenotype produce no corresponding change in the genotype: that is, the genotype remains unaffected or, if it is changed, the environmental effect is still not imitated by the action of the genes. This assumption has been tested in experiments on rat behaviour. McDougall, in a famous investigation, trained rats to escape from a water tank in which they were obliged to swim to the escape route. There were two ways out: one was a brightly lit alley, but entry into this was punished with an electric shock; the 'correct' way out was dimly lit. Successive generations of rats were trained in this way, and a progressive improvement in ability to perform the task followed. It seemed that the 'Lamarckian' transmission of 'acquired characters' had been experimentally demonstrated [204].

As usual with any pronouncement on this subject, a controversy of some violence arose, and McDougall's experiments were extensively criticized. An error of method, into which McDougall fell, was failure to run a control group: some of the rats in each generation should have been kept untrained; had he found an improvement in maze learning in this group also, no effect would have been attributable to the training. Munn [233, pp. 38–9] gives a more detailed critique. The most important criticisms are those of Agar and his colleagues, since they are supported by experimental evidence. These workers bred rats for twenty years, and trained them, in a manner similar to that of McDougall, for fifty generations. Again the rats showed a progressive improvement with time, at least during the first ten years. A control group, however, showed a similar improvement during the same period; later, both the experimental and the control rats underwent a decline in performance and then improved once again (figure 44). These fluctuations, which have

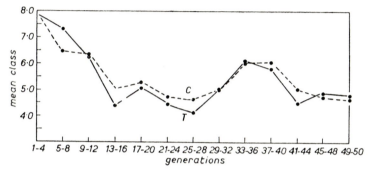

44. Test for 'Lamarckian' transmission. The curves represent mean ability in swimming a maze, by generation. The continuous line refers to rats whose parents had been trained in the maze, the broken line to control rats with untrained ancestors. *Both* lines show an improvement at first, indicating that the change cannot be attributed to the 'inheritance of acquired characters'. (After Agar, Drummond, Tiegs & Gunson [5].)

important implications for all experiments involving the study of an animal stock for many generations, are attributed to uncontrolled environmental, especially dietary, variations. The detailed reports of this research, maintained with such tenacity through depression and world war, show that the processes involved in this kind of learning are exceedingly involved; they are well worth careful study in the original [4, 5].

In the past it has usually been taken for granted that acquired characters, in the usual phrase, are inherited; even Darwin made this, for his day, conventional assumption. In some passages of his works he wrote of instincts as inherited memories [82]. He did, however, also propose that instincts were the products of natural selection [81, ch. 7]. A famous argument against Lamarckism, advanced by Darwin, is based on the non-appearance of the fixed action patterns of worker bees (and other social insects) in the fertile males or females. In this case there is no question of inherited memory, nor of loss by disuse. Recently, Waddington has demonstrated a process by which an apparently Lamarckian effect can be produced by selection in special circumstances [326]. Haldane & Spurway have discussed whether this genetical assimilation could bring about the replacement of a response, learnt by ancestors, by a stable fixed action pattern in the descendants. For this to happen it would be necessary for possession of a fixed pattern to confer an advantage over possession of the ability to learn. In their words, 'adaptedness would have to be made to confer more fitness than adaptability in postnatal life' [121]. The trend of mammalian evolution has of course been the opposite, that is, towards greater individual adaptability and reduced dependence on fixed patterns. In any case, we are left with no grounds for believing in any Lamarckian effect on behaviour.

5.4 'INSTINCT'

If then we may safely disregard Lamarckian hypotheses, it is appropriate to turn to Darwin (despite his own Lamarckian assumptions) when we consider the concept of instinct. Historically, the term 'instinct' has referred to rather mysterious abilities in man or other animals: its meaning, often vague, has been allied to that of 'intuition', and still is so allied in ordinary speech. One of Darwin's contributions to the study of behaviour was to encourage the rational study of the most complex and inexplicable activities of animals. Among these are the exquisitely patterned actions which enable some species to fit their lives with extraordinary precision to specific habitats or modes of existence. Rats themselves do not provide very good examples of this sort of thing: they do not make elaborate constructions like those of beavers, birds or bees; nor is their behaviour adjusted to any special source of food, or to any narrowly restricted range of environments. Not all rodents are so adaptable. Harris has shown that different species of *Peromyscus* respond differently to artificial environments with the visual properties of woodland or grassland respectively [127]. The success of rats seems to be largely a result of their adaptability to a remarkable variety of diets and circumstances. Perhaps this, together with their combination of exploratory behaviour and the avoidance of new objects, which has enabled them to achieve so much in human communities, is the best example of special behavioural 'adaptation' in rats.

For biologists, ever since Darwinism came to be accepted, all such abilities or adaptations are products of natural selection. If, within a population, there are two genetically determined varieties; and if one confers greater longevity or fertility than the other; then, over the generations, provided the environment remains sufficiently constant, the one will oust the other. The successful feature, in such a case, is

12. 'Abnormal' behaviour. A resident male wild rat (right) behaves momentarily towards a strange male as if the latter were a female—an interlude between conventional attacks. In the foreground a female retreats from the disturbance. (1/1250 sec.)

13. 'Abnormal' behaviour. A female wild rat (whose behaviour was atypical in several respects) behaves like a male making an ill-directed attempt at coitus with a normal female.

14. The adrenal cortex as an index of 'social stress' in wild rats. (a) (left) Part of a section of the gland of a control male. The heavy staining with sudan black indicates the presence of much cholesterol, a hormone precursor, in all layers of the cortex. The medulla, and the nuclei of the cortical cells, are unstained.

(b) (right) Similarly treated gland of a male which had died, though unwounded, after being subjected to attack for 14 hr. The adrenals, removed very shortly after death, had lost most of the cholesterol in the zonae fasciculata and reticularis, though the zona glomerulosa (at the top) had retained it. (From Barnett [22].)

100μ

said to confer greater fitness on those that display it. This principle, as Darwin proposed in chapter 7 of *On the Origin of Species*, applies to 'instincts' as much as to structures or other traits. The origin of elaborate behaviour patterns can thus be discussed, not as the work of a mysterious providence, but as the consequence of a process susceptible to rational analysis and open to experimental inquiry.

Darwin's discussion of this subject deals principally with 'instinctive behaviour', that is, with overt behaviour of the kind we have called 'fixed action patterns' and not with their internal causes. Today we find that in scientific writings on behaviour the word 'instinct' is used in at least two main ways. 'An instinct', in some writings, means a fixed pattern of behaviour, that is, a system of movements. This usage represents a drift of meaning from (i) an agency inside the animal to (ii) an activity evident on direct observation. But in other writings 'an instinct' still refers to an internal arrangement: it means, in fact, some neural, or neuro-endocrine organization which, when activated (for instance by a releasing stimulus), brings about a fixed action pattern.

This dual meaning of 'instinct' does not exhaust the possibilities for confusion. Haldane & Spurway, in the work quoted above, refer to the fact that Tryon's bright rats in the dark chose 'spatial hypotheses' such as right-hand turns before left-hand turns, while dull rats chose visual hypotheses, such as dark instead of light openings; and they add: 'What Krechevsky called a hypothesis might be called a type of instinctive behaviour' [121]. In this sense the 'type of instinctive behaviour' is not a fixed action pattern, as we have defined it, but the attainment of a system of relationships with the outside world which can be produced by a number of different patterns of movement. In this it resembles exploratory behaviour and the avoidance of new objects.

A still further source of confusion comes from psychoanalytic usage. Here the word *Trieb* (used by Freud and his followers writing in German) has usually been translated as 'instinct'. Etymologically and in meaning *Trieb* is allied to 'drive'. The term 'drive' is generally used, not to refer to patterns of movement, but to mean internal processes which make an animal active; sometimes an epithet is attached to 'drive', for instance 'hunger', and then the term refers to processes which make an animal active in some particular way. The behaviour may be stereotyped or variable: the specific character of the behaviour lies in the end or goal which it tends to achieve. 'Hunger drive' leads to eating. The researches described in § 3.4 are an example of an attempt to analyse certain 'drives' in physiological terms. *Trieb*, as used in psychoanalysis, has a meaning which is at least closely allied to that of 'drive' in this sense. It remains only to add that in ordinary speech, too, both 'drive' and 'instinct' may have (roughly) the same meaning as *Trieb*. It is not surprising that writings in this field have often been confused and confusing. This topic is taken further later on (§ 8).

Because the word 'instinct' has been used with such a variety of meanings, with much resulting confusion, it is not used in this book. This is a matter of *convenience*. It would be logically acceptable to announce a definition for this word (a *stipulative* definition), and to use this meaning throughout; but psychologically this procedure would be questionable. The term 'instinct' has different associations for different people; in some students of behaviour it arouses resistances which might preclude attention to the actual meaning of any passage in which it occurs. This is one reason why the term has been avoided. This does not of course at all entail neglect of the *facts* or *concepts* which are commonly discussed under the heading of 'instinct'. All that has been discarded is a particular *word*.

It is in fact still necessary to ask what is meant when, for instance, a writer gives an account of 'the instincts' displayed by the members of a taxon. Thorpe [312] gives the following list for birds and mammals:

nutrition	social relations
fighting	sleep
reproduction	care of body surface

Many comments could be made on this selection. Thorpe himself implies that the list is not complete, and that many questions could be asked about the different entries: for instance, to what extent are they related or interacting? But an examination of the list shows that it is heterogeneous as well as incomplete. 'Nutrition' refers to the fact that animals ingest a variety of materials, and sometimes perform complicated acts beforehand; some at least of this behaviour is essential for individual survival. 'Reproduction' refers to a great diversity of activities, not necessarily related to each other except by virtue of their survival value for the species; admittedly, sometimes the same hormone activates two or more of the behaviour patterns involved. Sleep is an inactivity. The incompleteness of the list is at once shown if one mentions, first exploratory behaviour or second, the maintenance of a particular skin temperature. Such a list, in fact, is no more than a rough grouping of the main goals that animals tend to achieve – and, indeed, must achieve if they and their species are to survive. The reader is reminded that by a goal is meant an activity (consummatory act) or a state (consummatory state) which ends a particular behaviour sequence. After eating, a rat turns to grooming or sleeping, and so on.

This sub-chapter shows, then, the range of topics often discussed under the heading of 'instinct'. The argument outlined here will be further supported in § 8.

5.5 SYNTHESIS

By way of conclusion there follow the principal features of complex stereotyped behaviour.

(i) The starting point of this chapter is the existence of fixed patterns of behaviour, that is, patterns of movement which are typical of a whole species or other taxon; another way of describing the standardized nature of these actions is to say that they are highly predictable. They are not sharply distinct from the acts commonly called reflex; but those most studied are much more complex and include many reflexes as components of the whole.

(ii) A fixed action pattern commonly differs from most reflex acts also because the readiness with which it can be evoked varies: the threshold is usually lowered with the length of the interval since it has last been performed.

(iii) The performance of a fixed action pattern thus depends on the internal state of the animal as well as an external stimulus or situation. Often, an internal state induces generalized movements, or appetitive behaviour (§ 2.2); these movements occur in the absence of a specific object, such as food, or of a particular kind of situation such as that provided by a nest; they end when a specific situation is achieved. This is called a consummatory state.

(iv) The two preceding generalizations do not hold for all stereotyped behaviour. There is no evidence that, in rats, being deprived of the opportunity to fight or to carry out amicable acts leads to an increased readiness to perform these actions, or to an increase in appetitive movements. Fighting occurs only in response to a particular kind of situation, one which is not in any sense 'sought': a group of siblings can live together without conflict.

(v) Fixed action patterns are not only themselves standardized: they are usually evoked by standard forms of stimulation, such as an odour, a visual pattern, a sound pattern or (at least in mammals) a combination of stimuli affecting more than one sense. A single odour, visual pattern and so on, having this effect, is a releasing stimulus. But here again there are exceptions: eating, for instance, as we saw in § 3, is in rats by no means dependent on a single stimulus pattern; further, just what stimuli do evoke feeding are determined in part by individual experience. Moreover, the response to a pattern, visual or auditory, is to some extent generalized: the different appearances of an object seen all evoke a response as does a given sound pattern even when presented in many different forms; this is due to learning (§ 7.3.1.3).

(vi) Non-specific features of the animal's surroundings may also influence the performance of fixed action patterns. For example, an adult male rat attacks other males only in a familiar area. Here again, learning (of topography) is involved.

(vii) The fact that behaviour patterns are often so highly standardized has led to the notion that they are 'innate'; but it is now evident that the term 'innate', unless it merely means 'stereotyped', is inappropriate. The question to ask is: how does the behaviour develop? The answer turns out to vary widely, even among the few examples which have been thoroughly studied. This is because there are several classes of agents which influence the development of behaviour; and each can influence different features of behaviour in different ways. The following list is slightly modified from one given by Hebb [136].

(a) Genetical: these are all the factors present at fertilization (§ 5.3.1)

(b) Prenatal: the effects of the uterine environment on later behaviour have been little discussed in this book; the most obviously likely sorts of effect are nutritional or toxic

(c) Postnatal chemical effects – again, nutritional or toxic

(d) Inevitable sensory effects: these include all the kinds of experience that every member of a species must have, such as those which (in a mammal) accompany feeding in infancy; they can be interfered with experimentally

(e) Variable sensory effects: these are special to the individual

(f) Injurious: these include wounding, infection and other 'stressors'

Whether a particular sort of behaviour is very stable or relatively labile depends predominantly on the extent to which it is influenced by agencies of classes (d) and (e).

(viii) No single scheme can accommodate the variety of the fixed action patterns even of a single species. This is not surprising: stereotyped behaviour, like everything else about an organism, is a product of natural selection. Genetical variation in fitness, which enables natural selection to take place, influences every bodily feature and process. Accordingly actions, all of which make an elegant contribution to the survival of the species, may be products of exceedingly diverse bodily processes: they may depend to different degrees on different sorts of sensory input and internal states; and they may be more or less labile in development. There is no simple general description of stereotyped behaviour to match the standardization we find in each single pattern.

A Taxonomy of Learning

... we are justified in raising the question whether the concept of learning or of memory embraces a unitary process which can be studied as a single problem, or whether it may not instead cover a great variety of phenomena having no common organic basis.

K. S. LASHLEY

6.1 INTRODUCTION

A naturalist, Bagnall-Oakley, has described some events which follow occupation of a starling (*Sturnus vulgaris*) roost at Egmere in East Anglia. Within a few weeks of the birds' arrival, large numbers of rats move into the earth banks nearby. The rats move about at night, and attack any bird that falls to the ground; a fluttering bird at once attracts a rush of rats from all quarters [16]. In each chapter so far reference has been made to such *changes in individual behaviour due to stimulation*. The phrase emphasized is one definition of the term 'learning' [312, p. 50]. In this sense it refers to a kind of overt behaviour. Another definition (already quoted in § 1) was given by a committee which debated the meanings of terms used in ethology: 'internal changes which manifest themselves as adaptive change in individual behaviour as a result of experience' [311]. Here the reference is not to behaviour but to processes which, in a mammal, go on principally in the forebrain. Some account of what little is known of the neurophysiology of behaviour is given later in § 9. In this and the next chapter we are concerned almost entirely with overt behaviour, and the first definition above applies to the term 'learning' as used in the chapter head. We shall see that the definition is inadequate; that the different sorts of behaviour it comprehends are very diverse; and that it is doubtful whether they all should be brought together under the one term.

One of the difficulties about studying learning is that by its very nature learned behaviour is so variable. We are accustomed in biology to variation between individuals (much of it of genetical origin) and we have methods for dealing with it; but the behaviour of most animals (including all mammals) is made more variable still because the central nervous system includes devices for altering behaviour in accordance with individual experience. To quote Thorpe: 'in the individual, self-preservation by means of adaptation to a changing and unstable environment is one of the most reliable criteria for distinguishing living from non-living' [312]. Such adaptation is seen in the hypertrophy of organs, for instance muscles or glands, resulting from use; but above all it is exemplified by changes in behaviour which tend to keep an animal alive or to enable it to reproduce. It is this feature which leads to the use of the word 'adaptive' in the second definition given above.

Since environments vary, the details of individual experience in the wild state are far from uniform even within a small population. Hence, as far as behaviour is concerned, the diversity already present at fertilization is magnified and made more complex by the learned adjustments of each individual. This is the basis of 'personality' – a phenomenon most obvious, and most highly developed, in man, but evident enough even among quite lowly mammals such as wild rats. If cannon balls or guided missiles or planets had personalities of their own, and their behaviour in response to repeated stimuli changed progressively, the subject of physics would present difficulties which, fortunately, do not actually exist. Physicists would be obliged to make observations on vast numbers of individual objects and to construct case histories of each: they would then perhaps be able to predict the behaviour of individual bodies (from previous observations of those bodies), but they would still be faced with the problem of grouping the individuals in classes concerning which more general predictions could be made.

This is the situation with which students of complex behaviour have to cope. Consider a superficially simple set of observations, such as those described in the first paragraph of this chapter. Or, better, we have more systematically reported observations such as those given in § 3.2.2. A group of wild rats occupies a neighbourhood. At one point in their range a pile of food is put down every evening. The rats gradually 'learn' to come to this point every twenty-four hours, at a fixed time, and to make a substantial meal. There are, however, considerable individual differences in the behaviour of the rats, even if we concern ourselves only with, say, adult males. How are we to investigate this sort of behaviour? Among the questions which may be asked are the following: (i) to what features of the situation is a rat responding during learning? (ii) are there different forms of learned behaviour (or can the behaviour be analyzed into distinct components)? (iii) what internal states influence learning? (iv) what is the neurological basis of learning? (v) how does the past history of the individual affect its learned behaviour? We have already seen, in § 5.3.1, that there is genetical variation in learning capacity, even among laboratory rats.

In this chapter we shall be concerned largely with the forms or components of learned behaviour: the analysis of complex sequences into elements such as habituation, trial-and-error behaviour and so on not only makes description easier but also helps in the design of experiments. A more detailed examination of the stimuli and other agents which influence learning is given in § 7. One category of learned behaviour, imprinting, is left until § 8.4.1.

6.2 'ASSOCIATIVE' LEARNING

6.2.1 *General*

A non-technical but agreeable classification of learned behaviour is into two classes: (i) simple; (ii) intelligent. We do not regard it as a sign of 'intelligence' if we become accustomed to a noise which at first disturbed us; or if we salivate at the sight of a lemon; or if we solve an irritating puzzle as a result of a series of 'random' movements. On the other hand, if we solve a problem by an insight into the relationships of its parts, or by putting together two hitherto quite separate items of knowledge, we call that intelligent. The first group, of simple learned responses, may be considered together under the heading of 'associative' behaviour: in each class of activities a particular response, or absence of response, gradually becomes associated with a specific situation. The second group is more difficult to define, to describe or to analyse; but, although it is not sharply distinct from the first, it is convenient to describe it separately.

6.2.2 *Conditional reflexes*

We begin, then, the systematic treatment of learned behaviour with an account of conditional reflexes (CRs). (The adjective 'conditioned', traditionally used in English, is a mistranslation from the Russian.) Alternative names for this sort of behaviour include 'classical conditioning' and 'conditional reflex type I'. The names arise because a different sort of behaviour, described in § 6.2.4 below, has often been called 'conditioning' too: this is 'trial-and-error', 'instrumental (or operant) conditioning' or 'CR type II'. In this book the term CR is used to mean 'classical conditional reflex', and that alone.

The researches on rats in which CRs have been studied have been well reviewed by Munn [233]. By contrast with studies of other forms of learned behaviour, the amount of these researches is small. Some workers have used the CR method to study sensory acuity. Cowles & Pennington (quoted by Munn) tried to establish the auditory threshold to certain sounds. If a rat has a shock applied to its tail, it squeaks; rats were trained to make this response to a tone which was sounded before the shock was applied. This is a typical example of the method and of a use to which it is often put. But the important studies have been not on rats but principally on dogs. The following account will in some ways be unfamiliar to readers of elementary texts; later, in § 7.2, there is an additional critique.

The CR is a kind of behaviour which is usually seen alone only in the laboratory. It depends on subjecting the animal to exceedingly restricted conditions. A particular response is selected by the experimenter: this may be a pattern of muscular contractions or a secretory activity: often it is something that can be studied quantitatively, such as a change in respiratory rate or heart rate. Whatever the response, it must, before the experiments begin, be educible by a specific stimulus. In the most famous of all such experiments, those of I. P. Pavlov (1899–1936) and his colleagues on dogs, this *unconditional stimulus* was the presence of food in the mouth; and the unconditional response was salivation. We have then the situation that a 'releasing stimulus' (if that term is preferred) regularly evokes a stereotyped response:

$$S_1 \rightarrow R_1$$

Typically, the response is one which seems not to be learnt: that is, once a particular stage in development has been reached the response appears without previous practice. For instance, food in the mouth causes salivation as soon as food is taken at all.

The preparation of a dog for CR experiments, described by Reese & Dykman [250] is exacting and laborious. Many months may be taken, during which the animal is habituated to the experimenters, the laboratory and the harness into which it has to be strapped for the experiments. The training itself consists of repeatedly applying a new stimulus, the *conditional stimulus*, at about the time at which the releasing stimulus is also applied – usually, a few seconds before. The conditional stimulus must be one which, applied alone, does not evoke the response to be studied, or anything like it. For instance, a sound or a light will serve if salivation is being observed. The method used by Pavlov can employ stimuli, conditional or unconditional, which act on any exteroceptor, and probably any internal sensory system too (cf. § 7.3.1.1). The conditional stimulus must produce *a* response, or at least a change in the animal's sense organs and nervous system. It is at first usually followed by a pricking of the ears, turning of the head or some other sign of 'attention'. Pavlov called this the 'what-is-it? reflex'.

The effects of the training are complex. One of them is that a response resembling to some extent the original response, comes to be evoked by the conditional stimulus:

$$S_2 \rightarrow R_2$$

In Pavlov's original experiments the formerly 'neutral' stimulus, usually a sound, now evoked salivation in the absence of the original releasing stimulus. A general and novel feature of this new response is that it is not performed with increasing vigour as the intensity of the stimulus is raised: on the contrary, there is an optimum intensity, and sufficient departure from this in either direction will prevent the response. Another feature of the new response, in the typical case, is that it is anticipatory, since the time-sequence of events is changed. When the experiments begin, the sequence is this: S_2 is followed by S_1 after, say, several seconds, and this in turn by salivation. (The delay may, however, be as long as several minutes.) After training the sequence is: S_2, followed by salivation after less than one second, (followed by S_1 or by nothing). Thus the new response occurs in advance of the time at which the original stimulus was applied. (This still holds when the delay is a long one: salivation occurs just before the time when the releasing stimulus is due.) A second effect of the experimental procedure is that the 'what-is-it?' response disappears. Other effects include changes in posture, in the heart and respiratory rates (figure 45) and possibly other features of behaviour and physiology. In particular, at first the releasing stimulus leads to increases in heart and respiratory rates proportional to the intensity of the stimulus; with training, these responses decline.

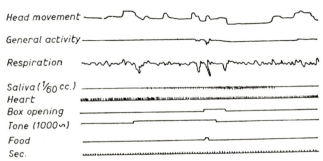

45. Conditional reflex in pig (*Sus*). At the sounding of a tone anticipatory salivation begins. Food (a biscuit) is dropped into the box, the pig noses the box open and eats the biscuit. Other changes are recorded. (After Liddell [198].)

It is a familiar fact of human behaviour that what has been learnt is often forgotten; and that a variety of influences, external and internal, determine the character and the rate of both learning and forgetting. These phenomena were studied by Pavlov; and, by using the technique of the CR, he presented them for the first time in a rigorous and quantitative form.

A first important finding was that the proposition, $S_2 \rightarrow R_2$, is misleading, if it is taken to imply that, of the total sensory input, only the conditional stimulus need be considered. This is shown in experiments which demonstrate what Pavlov called *external inhibition*. To be sure of establishing a CR reliably, it is essential to standardize the conditions of the experiment: the whole pattern of sensory input must be the same throughout the period of training and subsequent testing of the response. If, once training is complete, an unfamiliar stimulus is applied at the same time as the conditional stimulus, the response usually fails: the effect of the training is annulled. The animal is in fact affected by the situation generally: it is not like a machine set going by the pressing of a particular switch.

The second principle is a truism, but none the less important. The animal's internal state, or *motivation*, as we have seen in previous chapters, determines which ones, from among its repertoire of possible responses, can actually be evoked. If the salivary response to food is to be studied, the animal should have been deprived of food for some hours beforehand. The food thus constitutes a *reward*, or reinforcer, and presenting food is *reinforcement*.

The third principle is related to the second: the CR is lost if the conditional stimulus is repeatedly presented, over a short period ('massed trials'), without the releasing stimulus. This has been called 'internal inhibition', but is now more usually (but less appropriately) named *extinction*. This sort of thing is, again, a matter of everyday experience. It is also to be expected if one considers the function of the CR, since it is no use salivating if there is no food. Extinction, then, is a corollary of the relationship between the animal's internal state and the releasing (unconditional) stimulus. The latter must be provided by an object which enables the animal to alter its internal state: it must enable the animal to perform a consummatory act or to achieve a consummatory state.

G. B. Shaw once complained that the study of CRs yielded nothing but statements of the obvious. That this is not true appears when extinction is examined further. Suppose that a CR is extinguished, by repeated presentation of the conditional stimulus without reinforcement, on one day. If, on the next day, the conditional stimulus is presented again, the response will reappear. The animal has, so to speak, forgotten that the response to the conditional stimulus is no longer rewarded. This *spontaneous recovery* suggests that extinction is not merely due to a reversal of whatever neural changes have taken place during conditioning: evidently, the CR is due to one set of neural processes and extinction to another. (Hence follows the superiority of the term 'internal inhibition' for 'extinction'.) These processes may be compared with the excitation and inhibition of neurons familiar from the work of neurophysiologists.

Whatever the neural basis of inhibition, it is found to be related to excitation in a complex way. (These terms are here used to describe behaviour, not neural processes.) First, inhibition can itself be nullified by external stimuli. If a novel stimulus is applied at a time when, as a result of its training, a dog is withholding saliva, the dog may suddenly salivate copiously. This is easily shown when an animal has been trained to delay salivation during a long interval between the conditional stimulus and the unconditional stimulus. The phenomenon is called *disinhibition*; it parallels external inhibition of a positive response. In this respect, in fact, conditional excitation and conditional inhibition resemble each other: both can be annulled by an 'irrelevant' stimulus.

Inactivity can be induced by training just as easily as activity. The clearest example is sleep, which was the subject of a number of studies by Pavlov himself. In fact, all training to perform a specific act also involves training not to respond in other ways: this was exemplified above when the loss of the 'what-is-it? reflex' was described. This phenomenon, *habituation*, is further discussed in the next section.

There is another aspect of the interaction between excitation and inhibition. Suppose that a dog has been trained to salivate at the sound of a metronome, and that it has been trained also *not* to salivate at the shining of a light; then, if the light is shone, and the metronome switched on immediately afterwards, more saliva is secreted than would have resulted from the metronome alone: the response may even be doubled. This is *positive induction*. The complementary experiment is to switch on the metronome and follow it at once with the light; this is followed by an inhibition which is exceptionally resistant to disinhibition. So we have negative induction.

Apart from these many features of the simple CR, Pavlov's method allowed study of *second order conditioning*, recently reviewed by Mowrer [231]. Consider a dog which has been trained to salivate to a flashing light. The light can be presented with a new stimulus, such as the sound of a metronome; and the sound can in this way be made to evoke salivation even though it has never been presented in conjunction with the unconditional stimulus (that is, food). In these conditions the original stimulus (the light) comes to act as if it were a reward. The same applies to a situation in which a dog, trained to salivate to a light, is given the opportunity to switch the light on himself: he will learn to do so, just as though the light itself were a reward. Mowrer writes:

> the dog learns a new bit of behavior, not because that behavior produces (is rewarded by) food, but because it produces merely a sign, or *promise*, of food.

The systematic study of conditional reflexes, then, reveals a number of regular, predictable but complex relationships involving the animal's internal state, external circumstances and previous training. Of the concepts derived from, or used in, this study, those of excitation, inhibition, anticipation, motivation and reinforcement are particularly important. They will be discussed further in § 7.

6.2.3 *Habituation*

Meanwhile, it is necessary to say something more about inhibition, using the term in a behavioural sense to refer to a decrement in activity. In § 2.2.4 we saw that novelty or changing stimulation arouses a rat's attention; correspondingly, repeated exposure to the same stimulus is liable to result in a decline in the response originally evoked; this holds provided the response is not reinforced, that is, does not lead to achievement of a consummatory act or situation. Another aspect of this phenomenon was described in § 2.3: the avoidance of a new object displayed by wild rats declines with exposure to the object.

All these are examples of *habituation*, if we define that term as the waning of a response as a result of repeated stimulation. (The term 'habituation' is sometimes given a more restricted meaning; the definition given is the one used in this book.) Habituation, so defined, includes 'negative adaptation'; this is a term usually applied to the waning of a simple reflex and other fairly simple responses. For example, a rat jumps if it hears a sudden high-pitched noise. This startle response has been studied by Prosser & Hunter: they evoked it by means of a click, sounded at intervals of fifteen seconds in a sound-proof room; the response was measured by means of electrodes in the leg muscles, usually the gastrocnemius. After thirty clicks the response declined; during another thirty-five it was irregular; eventually it disappeared [246].

This sort of gradual loss of response is found throughout the animal kingdom, especially in relation to movements of withdrawal and other protective responses such as the eye-blink. Habituation of these movements occurs when a stimulus is repeated, yet not followed by injury or further disturbance. The value of this type of learning is obvious. Habituation is also a necessary component of more complex activities. When a wild rat first encounters food in an unfamiliar place it must lose both its avoidance behaviour (§ 2.3) and its tendency to explore (§ 2.2), before it can form a regular habit of visiting the food with a minimum of effort: it has to learn to ignore features of the situation other than the food.

A question sometimes asked is: how do we distinguish 'true' habituation from fatigue? To answer it, the question must be re-formulated. The definition of the term 'habituation' given above includes phenomena which would commonly be called examples or consequences of fatigue; if this is disliked, the definition must be altered. However, as Haldane [118] has pointed out, students of behaviour are sometimes inclined to call the waning of a response 'habituation' if they approve of it, 'fatigue' if they do not. It is more useful to consider the question: are there different kinds of habituation (as defined); and if so, is one kind a result, say, of the accumulation in muscles of metabolic waste products, another a consequence of changes localized in sense organs and yet a third due to changes in the central nervous system?

Sensory accommodation is a familiar phenomenon, described in textbooks of physiology. Recovery from it is rapid. If it were not rapid, the sense organs would be out of action most of the time. However, we cannot by any means assume that habituation of very short duration is always due to changes in sense organs. Thorpe, in a discussion covering the whole of the animal kingdom, shows that some very brief response decrements are due to central changes [312]. Recovery from the sort of habituation exemplified above takes longer: while a sense organ often recovers in a few seconds, habituation of the startle reflex to the criterion of no response to five successive stimuli is usually followed by spontaneous recovery only after about twenty minutes [246]. In other instances the delay is far longer.

What we can say with confidence is that prolonged habituation is not due to the accumulation of metabolites in muscles, but is due to processes in the central nervous system. If repeated stimulation has abolished a rat's startle reflex, an additional disturbance such as a flashing light at once restores it in full. This shows that the loss of the reflex is not accompanied by any important decline in muscular responsiveness:

the change must be central and indeed probably in the forebrain and reticular formation, where paths from the ears and eyes meet; from the forebrain the pyramidal tracts pass to the motor neurons of the spinal cord. The fact that habituation can itself be inhibited by an interfering stimulus shows that the underlying processes in the brain are unlikely to be simple.

There is much more that could be said about habituation, but fortunately it has been very fully discussed by Thorpe [312], to whom reference should be made for further details. He makes it clear, as Hinde has in experimental studies [146], that habituation as defined here covers a number of very different processes.

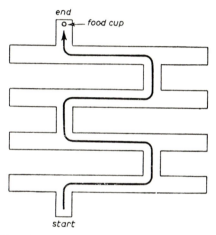

46. Plan of a typical enclosed maze. (From Tsang [321].)

6.2.4 Trial-and-error behaviour

Much of the great volume of researches on simple learned behaviour in mammals has been described under the heading of 'conditioning'; but usually both the methods used and the behaviour studied have differed in important ways from those of Pavlov. We have seen that the names given to the behaviour (or to the underlying internal processes) include trial-and-error, operant conditioning, instrumental conditioning and CR type II.

An unsophisticated example was described by Small, at the beginning of this century, from his pioneering studies of laboratory rats. Each rat was made hungry and then had a box with a hole in it put into its cage; the hole was covered with sawdust. The box, which was already familiar to the rat, had food in it. The rat eventually found its way into the box, as a result of what Small described as random scratching around in the sawdust. For one rat the time taken on the first presentation was ninety minutes, but on the thirteenth occasion it was only half a minute [292].

Small's experiments involved the rat in learning to find its way to a goal. This holds also for the many experiments in which rats are required to run a maze (figure 46). Examples of the use of this technique have been given in §§ 2 and 3. Its history, and the variety of mazes used, have been summarized by Munn [233] and by Mowrer [230]. At first, mazes were elaborate systems of pathways, like the famous one at Hampton Court where people can spend a sunny afternoon getting lost. However, the experimenters too found themselves at a loss, when they were obliged to interpret the results of running rats in mazes with many choice points. Eventually it became usual, for many purposes, to use T -or Y-mazes with only one choice point (figure 6, page 20). However many branchings there are, a maze may be a system of walled passages, of raised runways, or of canals through which the animal has to swim. They may be variously illuminated or dark; in a featureless room or not; and in a sound-proof room or a noisy one. Whatever the details the rat is faced with one or more choices; in a complex maze of which it has had no previous experience it gradually eliminates the alternatives, until the path from the start to the goal is run rapidly and without error or hesitation. During the process of learning, a rat may reach a stage when it makes no errors at a particular point, but still pauses and makes incipient movements, first in one direction, then in the other. Muenzinger, in 1938, called this behaviour 'vicarious trial and error', or VTE. It is reminiscent of the 'intention movements' which have been much studied in birds, though these are preliminaries to performance of stereotyped acts and do not represent a stage in learning. VTE will be mentioned further below, in § 6.4.

Rats are well equipped to make their way through systems of tunnels or passages. Their capacity for topographical learning is considerable (§ 2.4.1). But they also have some powers of manipulation, used in feeding (§ 3.2.3) and in nest-building (§ 5.1.3). This sort of behaviour too may be evoked in trial-and-error situations. Tolman has discussed experiments in which rats learned to get hold of food by pulling a string [317, page 130]. Rats are allowed to feed from a pan, which on later presentations is slowly drawn away out of reach; as the pan is removed the rat seizes it and drags it back. Later, the only way in which the rat can get hold of the pan is by pulling on a string attached to it; it soon encounters the string while scrabbling for the pan, and quickly learns to pull the string, usually with its fore-feet but sometimes with its teeth.

Eventually, with the pan out of reach from the start of the experiment, the rat will readily pull it in with the string.

However, the most popular means of studying manipulatory trial-and-error behaviour is the puzzle-box. That most used for rats was invented by Skinner [290]. The Skinner box is illustrated in figure 8, (page 26). Its essential feature is a lever or similar gadget which, suitably moved, releases a pellet of food, a drop of water or other fluid or, in some experiments, switches a light on or a current off. There may be two levers, and many other variations. An inexperienced rat in such a box makes a great variety of movements. Eventually, if it is suitably motivated, and if moving the lever produces a suitable reward, it learns to press the lever without fuss.

The Skinner box may be arranged so that it presents a needed object; but the experimenter may also use the avoidance of, for instance, electric shock as the 'reward'. Other apparatus too has been used for this second purpose. Warner used a cage in which a shock could be administered through a floor grid. The shock could be escaped by surmounting a small barrier and entering another compartment. The shock was preceded by the sound of a buzzer, which was already known to produce no disturbance by itself. On their first experience of the shock the rats made violent movements of no specific direction – a rather frantic example of appetitive behaviour; but quite quickly the rats learned, on hearing the buzzer, to cross the barrier with economy of effort and smoothness of movement [327]. Yet another method, little used for rats, involves fastening the animal in a harness which severely restricts its movements. A shock is applied to some part of the body, but the animal can avoid the shock by making a specific movement, say, of the leg. If the shock is regularly preceded by a neutral stimulus, such as a sound or a light, the animal learns to avoid the shock altogether by making the necessary movement in advance. Again, in the early stages, many inappropriate movements are made.

The behaviour described in the preceding paragraphs is, in its most important features, much less diverse than the methods designed to study it. We must now examine these features, and we must distinguish the behaviour from the CR.

In the CR the response (R_2) is produced because a stimulus, which already evokes something like the response (R_1), is presented. This is, in a sense, a more efficient method of learning than trial-and-error, since in the latter a vast number of inappropriate movements have to be made, and eventually discarded, during learning. However, trial-and-error permits a much greater degree of novelty in the response, provided that the conditions allow the animal freedom of movement. The ultimate pattern is then an economical system of movements which, as a whole, is new. This final performance, whether it is a leap over a barrier or pressing a lever, is what remains from the original, less directed, appetitive behaviour. The whole pattern of movements may *include* CRs; for instance, finding food in a maze, or getting it in a puzzle box, must involve many proprioceptive and other reflexes and end with reflex salivation.

An important difference from the CR alone is in the time relationship with the reinforcing stimulus. At all stages during training the response which is eventually learned must be made before the reinforcement: the latter cannot occur until the response is made. Another way of putting this is to say that the response is *instrumental* in bringing the animal to its 'reward'. In the Skinner box, for example, the food pellet arrives *only after* the lever has been pressed. As a corollary of this, it is sometimes said that trial-and-error behaviour differs from the CR in altering the circumstances of the animal: once the animal hits on the right response it is enabled to make a specific change in its environment. However, this distinction is not a wholly valid one: for instance, the saliva secreted by the Pavlovian dog dissolves some of the food and makes it more digestible [135].

The most conspicuous feature of trial-and-error behaviour is of course the appetitive behaviour from which its name is taken. This is sometimes said to consist of 'random' movements. If it is used strictly, the term 'random' is reserved for circumstances in which no one movement is more probable than any other. It represents a statistical concept. But we saw in § 2 that a rat's variable movements are not random. They may be difficult to predict in detail, but some valid general statements can be made about them. For instance, wild rats are unlikely to approach a completely unfamiliar object in a familiar area; and all rats (with that one qualification) tend to explore and to sample the unfamiliar in preference to the familiar. The phenomenon of spontaneous alternation, described in § 2.2.3, illustrates the non-random character of a rat's movements in a particularly simple case. There is, however, another reason why randomness cannot be expected: it is that in most circumstances previous learning will influence the animal's choice of actions.

The description of trial-and-error behaviour may then be summarized as follows. (i) the learned response is a new one, at least in relation to the stimulus pattern which evokes it; the response is crystallized out of the multifarious actions which constitute appetitive behaviour. In this respect the learned response is unlike the CR. (ii) The difference from the CR holds for the complete pattern of the response; but this does not preclude the occurrence of CRs as *components* of the response. Indeed, both CRs and habituation constitute components of learning by trial-and-error. (iii) The factors which influence trial-and-error learning resemble those which control the CR: phenomena such as inhibition and the effects of reinforcement and of motivation can be observed in both kinds of learning. This is further exemplified in § 7. (iv) Although in this chapter trial-and-error behaviour has been labelled 'simple', it may well be doubted whether this is justifiable; the behaviour is indeed not sharply separable from that described in the next section.

6.3 'INSIGHT' BEHAVIOUR

The preceding account of 'simple' learning, though itself much simplified, shows that *all* learned behaviour, at least in a mammal, presents formidable problems, both of description and of explanation. With insight behaviour we come to a kind of learning in which these problems are still more severe. Several formal definitions for this term have been proposed. One, slightly shortened, is: the solution of a problem by a sudden reorganization of behaviour. Another is: a gross difference in behaviour on successive presentations of a problem, when the behaviour on the second occasion is judged by the experimenter to be efficient. This second definition, based on one by Verplanck [323], is perhaps unnecessarily cynical. For this book, the first definition is used, despite its vagueness.

In any case, like many other definitions – however carefully constructed – both are less informative than a full description of actual behaviour. As with trial-and-error, we are dealing with problem-solving. An animal is faced with a situation in which, we know, it will carry out some act, or achieve some end-state, if it can. The goal, as usual, is determined by the animal's internal state, which may be hunger, cold, sexual deprivation or some other state of need. To reach its goal, the animal must perform a series of movements which it has not performed before; even though it has already employed the component movements, it has never before carried out the total pattern. For most species in which this behaviour has been studied, the problem has been to find a way from one place to another; but for some, especially the Anthropoidea, solving the puzzle has required manipulation.

We have seen how an animal may solve a problem by a series of ill-directed movements from which the appropriate ones are gradually selected; and that these movements are not random in the strict sense.

One factor which reduces randomness is past experience; in problem-solving, an animal's performance may be much or little influenced in this way. In an extreme case an animal may perform the appropriate act without delay, error or unnecessary exertion on its first exposure to the problem. When it does this, it is of course making use of previous experience – experience, that is, of different elements which now make up the total situation; its behaviour, in such a case, depends in part on the exploratory learning that has taken place on earlier occasions, and not on any overt trial-and-error.

The traditional examples of this insight behaviour are those in which a monkey or ape employs tools, or manipulates a puzzle. Rats, although they do handle objects, are less dexterous; the problems which best allow them to display insight behaviour are of the kind presented by mazes. Figure 47 opposite illustrates the principle of a series of classical studies by Maier. Each rat is allowed to learn different parts of a maze on different occasions, and under different motivations. As usual, the rat is hungry or thirsty, and finds its way to a goal-box containing food or water. While one part of the maze is being learned, the others are blocked. Each part has a different kind of flooring; this makes use of the highly developed sensitivity of the skin of the rat's toes. In the test of 'reasoning' (as Maier called it) the rat is faced with a new situation: it is hungry; it is put at a starting point from which it can go left or right; both these paths have previously, on different occasions, led it to water, but from only one of the water points is it possible to go further to food. Maier found that rats chose the correct route more often than the wrong one [208]. In doing so, they made use of separate previous experiences, to make the correct movements on the first encounter with the problem.

On our definition, Maier's rats displayed insight behaviour. However, giving the behaviour this name is not an explanation. Some of the problems of explanation that arise with this sort of learning can be illustrated from another famous series of experiments. Tolman & Honzik, and later others, used a design of maze also shown opposite. Rats first experience the maze with all paths open, and learn to take the direct route to the goal. When this path is blocked at A, they learn to take the shorter of the two remaining paths. The test of insight behaviour is to block the direct route at B, and so to leave only the longest route open. In the original study, fourteen out of fifteen rats took the longest, but only open, path [318]. Later, however, Dove & Thompson repeated this experiment with some modifications and with equivocal results: most of the rats failed on the first trial, but got the answer right on the second or third trials, so suggesting rapid learning by trial-and-error.

In further experiments by these authors the blocks were made conspicuous by means of painted lines which could be seen from the choice point. In these conditions no animal chose the long route on any of the three test runs. Evidently it was the fact of being checked at the particular point at which block B was put, during a run on the straight path, which led to the choice of the long path [88].

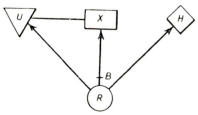

47. Experiment on insight learning: the principle of Maier's experiments on 'reasoning' in rats. A rat learns different parts of the apparatus on different occasions and later makes use of the separate experiences to solve a new problem. See text. (After Hull [156].)

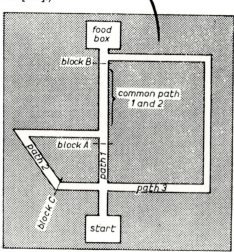

48. Experiment on insight learning: the modification by Dove & Thompson [88] of the Tolman-Honzik maze. See text.

The most obvious implication of these and other experiments (more fully reviewed by Anthony [11]) is that the conditions in which rats will display insight behaviour in an unequivocal way are narrowly limited. But this conclusion must be qualified. The experiments of Maier and of Tolman & Honzik were designed to show insight behaviour in a form in which it could not be confused with trial-and-error; but much of the behaviour which contains an 'insightful' element also involves some trial-and-error. In trial-and-error learning in which previous experience plays at most a very small part, the curve of learning takes a form illustrated in figure 54, page 168. This is in contrast to what happens when exploratory learning is effective: the response is acquired much more quickly – sometimes rather suddenly; this is exemplified in the second group of experiments by Dove & Thompson mentioned in the preceding paragraph.

Sudden changes in behaviour during the learning of a task are indeed general, at least in mammals. They have been much studied in man. A peculiar kind of example was first adequately described in rats, by Krechevsky. The form of the experiments was to give hungry rats, on successive occasions, a choice of two doors; the doors were distinguished by being painted white and black respectively. The white one was sometimes on the left, sometimes on the right. Always, one door led to food, while the other was locked. In human terms the problem was to decide whether colour or position marked the presence of food. In these circumstances a rat does not proceed at random: usually it persistently tries the door on one side, for several trials, then the other side; or it may alternate regularly, or choose always one colour, regardless of side. Eventually it settles down to whichever choice has consistently led to food. Krechevsky called this behaviour the formation of 'hypotheses', by analogy with human behaviour in similar circumstances [179]. The phenomenon of vicarious trial and error, mentioned in § 6.2.3 above, may be regarded as analogous to hypothesis formation; but VTE does not involve a complete movement – only an abbreviated one. It is possible that, in an extreme case, a process occurs in the brain, similar to that which brings about VTE behaviour but leading to no overt act. In human, subjective terms, this would be thinking, or forming hypotheses in the everyday sense of that expression.

We may now examine further what is meant by the phrase used above: 'behaviour which contains an "insightful" element'. This expression, like our definition of 'insight behaviour' (page 147) is vague. The main reason for this imprecision is evident enough: the behaviour we are concerned with is difficult to describe and classify (let alone explain) in a few words. One source of the difficulty is that we have distinguished two kinds of behaviour which are in fact not clearly demarcated. The first is trial-and-error behaviour. An inexperienced animal in a puzzle box provides the standard example: the appetitive behaviour is at first virtually undirected and, if not random in the rigorous sense, is unrelated to the needs of the situation. The second is insight behaviour. In an unequivocal example of this the solution of the problem is sudden, smooth and novel; one gets the impression that, by contrast with trial and-error, the solution *precedes* the movement which brings the animal to its goal: it is as if the animal has thought out the answer before moving, perhaps by an *internal* process of trial-and-error.

The difficulty arises when one examines the behaviour which lies between these extremes. The facts of 'hypothesis formation', exploratory learning and so forth, displayed in maze learning or similar tasks, intrude an apparently alien element into situations in which plain trial-and-error might be expected. The same applies to 'learning set', which is discussed later, in § 7.4. It follows that we must expect, in any example of complex learned behaviour, features both of trial-and-error and of 'insight'. This may be inconvenient; but it serves to remind us that natural phenomena are always more subtle and multiplex than the words we use to describe them.

6.4 THE QUESTION OF 'INTELLIGENCE'

In § 6.2.1 it was suggested that insight behaviour, unlike 'conditioning' and trial-and-error, would ordinarily be called 'intelligent'. It might be thought that the term 'intelligence', like 'instinct' and 'drive', has been used so vaguely and so variously that it would be better to avoid it. But, even if this is granted, it does not release us from the obligation to master the facts subsumed under the name. Here, by 'intelligence' is meant simply 'learning ability' – an expression which covers a mass of diverse phenomena.

At present the relevant facts concern for the most part the *problems* of testing intelligence. A first difficulty is the many agencies which influence intelligence. Early work on differences among rats in the ability to learn involved the selection, from a mixed stock, of genetically different strains. Some of the results have been described in § 5.3.1. The criterion of 'brightness' or 'dullness' was the score in maze learning. But the scores so derived reflect a complex of factors, ranging from the degree of need (for instance) for food at the time of the experiment, through early experience or training, to genetical constitution. All three of these could influence features of behaviour, such as exploratory tendencies or the avoidance of objects, which might affect learning ability.

A second difficulty is that of devising a scale of intelligence. When one wishes to compare the intellectual abilities of human groups or individuals, one sets a series of problems of increasing difficulty. Difficulty is assessed by the proportion of individuals capable of solving it, either in the population studied or in a standard population. Some problems must be soluble by almost all, some by hardly any individuals. With such a procedure, the criterion is a function both of the predilections of the person designing the tests and of the population chosen as a standard. There is no independent scale of intelligence, as there is, say, of weight-lifting ability.

A third difficulty, related to the first, is that of disentangling 'pure intelligence' from other qualities which influence its expression. A problem must present a goal, and a goal necessarily implies motivation; that is, the subject must be in a state which accords with the 'reward' offered on solving the problem. (This is still true even if the reward is only the actual operation of solving the problem, as further discussed in § 8.4.2.) If one is testing 'intelligence' one wishes to keep the factor of motivation constant. This is because it is commonly assumed that there is a quality, whether it is called 'intelligence' or 'IQ' or something else, which is independent of motivation. The extent to which this assumption is justifiable is, however, an open question.

We will now consider experiments which have some bearing on rat intelligence. A number of investigators have studied the ability of rats to learn a maze in which the correct route requires 'double alternation' of turns [reviewed by Munn, 233, p. 283 e.s.]; that is, the rat has to go right, right, left, left, right, right, and so on. If the conditions are carefully arranged so that the rat cannot distinguish different points in the maze by means of 'irrelevant' cues, such as different echoes from below the runway, it usually fails to get beyond r, r, l, l. There are, however, some examples of a probable learning of double alternation in some form. Evidently, this sort of task is on the borderline of ability of laboratory rats. Wild rats might find it easier, but we have no information on that point.

Another kind of task, also on the borderline of rat abilities, is the 'oddity problem'. The animal is required to approach one of three objects or patterns; the correct one is that which differs from the other two. Monkeys can certainly learn to do this, but it was for some time thought that it was beyond a rat's powers. However, in 1953 Wodinsky & Bitterman published the results of experiments on rats which had previously had much experience of problems in which they were required to jump towards one of three windows. (Figure 36, page 108, illustrates the sort of apparatus used, but shows only two windows.) The rats had also been trained to switch readily from (i) choosing a black card in preference to either of two white cards to (ii) choosing a white card in preference to two black ones. When, after this, the problem was to choose always the odd card (whether white or black), the rats quickly learned to do so. Other rats learned comparable tasks [331].

This work is significant not only for the problem of assessing learning ability, but also in showing how important is previous experience of problem solving. This is the phenomenon of 'learning set', or the acquirement of learning ability as a result of previous experience of comparable problems (cf. § 7.4). Petrinovich & Bolles have given another example of how previous training influences the 'intelligence' with which a task is performed. Rats were *trained* to alternate between left and right in a T-maze; after this, they continued to alternate to a significant extent even when the interval between successive trials was increased to several hours, although untrained they would then have displayed no alternation. In this case memory or retention was greatly improved by training [242].

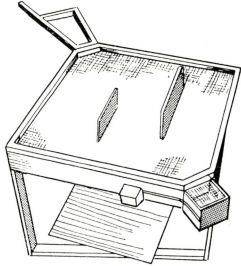

49. 'Intelligence test' for rats: the Hebb-Williams apparatus as modified by Rabinowitch & Rosvold [249]. See text.

It was during an inquiry into the effects of early experience that Hebb & Williams developed an 'intelligence test' for rats [138]. They were testing learning ability by training rats to find food in one of four containers in an open field apparatus, and this procedure revealed no difference between two groups of rats. The task took a long time to learn and, in Hebb's words, presumably should 'be classed as rote learning rather than insightful' [133]. Another test, however, showed a clear difference. The new method has been revised by Rabinowitch & Rosvold; their version is illustrated in figures 49 and 50. A series of tests is set, each in the same apparatus. Each item is a detour or *Umweg* problem which can be quite quickly mastered by a rat. Each rat has to be accustomed to the apparatus, to being handled and to eating in the food box; when this has been done the rat is given preliminary runs which are not used in scoring. The period of fasting is standardized, and so are all surrounding circumstances. The method is reliable, in the sense that re-tests give closely similar scores to original tests. It could undoubtedly be used to compare the ability to cope with

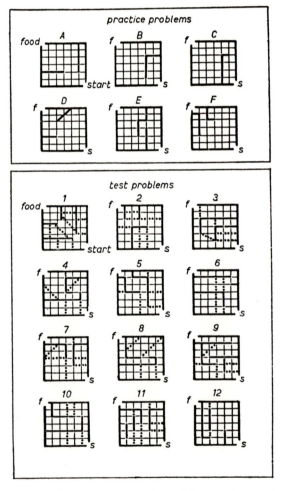

50. 'Intelligence test' for rats: the floor plans of the practice problems and test problems used in the Hebb-Williams apparatus shown in the previous figure. (After Rabinowitch & Rosvold [249].)

this sort of problem not only of different groups of rats but also of different species.

Nevertheless, it is not easy to interpret the results of comparing the abilities of different species. Lashley long ago pointed out that in a simple maze a rat performs as well (that is, as 'intelligently') as a man [185]. This point has been developed by Hebb, in experiments which show that for the learning of simple tasks rats may even be at an advantage over 'higher' mammals. He reared rats in the dark and tested

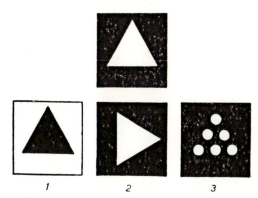

51. A test of 'intelligence'. A rat trained to respond to the top diagram makes random responses to any of the others. A chimpanzee (*Pan*) responds to diagrams I and 2. A child (*Homo*) of two years recognizes all three of the lower diagrams as triangles. (From Hebb [in 260].)

their ability to learn to discriminate a pattern of vertical bars from one of horizontal bars; they took six times as many trials as rats reared normally; to learn to distinguish a triangle standing on its base from one standing on an apex took twice as many trials. But – and this is the important point – within an hour of being put in the light the behaviour of the experimental rats was normal: that is, it could not be distinguished from that of the controls. A chimpanzee, and still more a man, would take far longer [133]. This is evidently a reflexion of their much larger mass of brain tissue – a tissue which confers on them the ability to learn tasks far beyond a rat's compass.

In general, 'intelligence tests' have to be fitted both to the species and to the particular purpose of the inquiry. Intelligence is not a simple, measurable quantity like body weight, even if we ignore the complicating factor of motivation. It is not enough to speak of the ability to learn: one has to specify what is to be learnt, just as with human beings. We have seen that one kind of maze, with few choice points, may be learnt as readily by a rat as by a monkey; another sort of task may never be mastered by rat or monkey. Between these extremes is a range of problems; some of these can be solved by monkeys but not by rats, but only (perhaps) by virtue of the monkey's possession of mobile fingers; yet others can be solved by both, though perhaps with more mistakes on the part of the rats. Small, in 1899, found that rats easily solved a problem which involved digging: they are burrowing animals, and this activity was within their standard ('innate') repertoire. But a task which required the tearing away of a piece of paper took several days to learn, for tearing is less commonly done by rats; but, once it had been done for the first time, learning was rapid [291].

The differences in learning capacities of different species are, however, likely often to be more subtle than those arising from their distinctive motor patterns. Fields and his colleagues (quoted by Hebb) studied the process of learning to respond to a triangle (figure 51). In a rat this evidently involves cerebral activities different from those in a chimpanzee (*Pan*); and the latter differs in turn from a two-year-old child (*Homo*) [in 125].

In fact the question of 'intelligence' forces one to realize once again that learning or learned behaviour is diverse, multifarious and complex. It is, however, possible to say more about the factors which influence learned behaviour, especially of the simpler varieties. This is the main subject of the next chapter.

7

Analysis of Learning

Learning psychologists are still groping with methodological . . . problems that they hope will lead to a major breakthrough in understanding behavior. Some believe a breakthrough is now occurring. But nobody is willing to recognize anybody's breakthrough but his own.

H. H. KENDLER

7.1 INTRODUCTION

The preceding chapter was concerned primarily with description: it gives the main categories of learned behaviour as they are often listed, although these categories are better regarded as components of most learned behaviour. This classification provides a convenient frame of reference for the ordering of many facts, but it could be misleading if it were taken for more than a provisional scheme. When more is known about the neural basis of learned behaviour we shall, perhaps, have to describe it very differently. As it is, the categories of CR, habituation, trial-and-error and insight behaviour reflect different experimental procedures as much as the actual facts of animal behaviour.

However, in the absence of a neurology of learning, we have to make what we can of behaviour as we see it. A major portion of this chapter is devoted to attempts to give an orderly analysis of trial-and-error behaviour (instrumental or operant conditioning). But first we have to dispose of a tiresome obstacle to understanding behaviour: this is the garbled account of the work of Pavlov and his followers still found in many elementary texts.

7.2 THE 'CONDITIONED-REFLEX' MYTH

A myth is a fictitious narrative embodying popular ideas on natural phenomena. This adaptation of an entry in the *Oxford English Dictionary* is an adequate description of many textbook accounts of conditional reflexes, though such accounts have long been discarded by most ethologists and psychologists. To dispose of this myth we must first restate some features of the CR.

(i) The CR is *not* the same as the original, unconditional act, although it is often represented as being so: (*a*) it is usually anticipatory; (*b*) the response chosen for study, for instance salivation, is usually only a part of the unconditional response to the releasing stimulus; (*c*) the chosen response is also only a part of the total behaviour of the trained animal, since during the experiment the animal may yawn, stamp or pant, its heart-rate may increase and many other internal changes occur.

(ii) The CR, in the sense of a particular response (such as squeaking or salivation) which has come to be evoked by a new stimulus, is only one sort of component of complex learning: it can by no possible means be stretched to cover all learned behaviour. The procedure invented by Pavlov reduces the animal's activity to a minimum. Restraint in a harness precludes virtually all appetitive behaviour. If an animal is trained in less confined conditions the complexity of what is going on during 'conditioning' becomes more obvious: in addition to the hidden changes in heart and respiratory rates, a variety of movements are made; behaviour in fact takes a familiar appearance, namely, that of an animal which (in colloquial terms) is expecting something to happen and awaiting the event with some degree of anxiety or at least interest. (These anthropomorphic terms should, perhaps, be put in quotation marks; but they are used here to describe shortly a kind of behaviour known to everyone who has kept a pet or has watched laboratory mammals informally.)

(iii) We have seen that behaviour in which the appetitive element is very obvious, that is, trial-and-error, has been called 'CR type II'. It may therefore be said, in response to the last two paragraphs, that at least all learned behaviour can be comprehended under the CR heading if one includes type II also. This may well be the case: but trial-and-error behaviour includes not only an appetitive component but also, to varying degrees, an element of insight behaviour, as defined; that is, it includes that ability to make use of two or more separate past experiences, to give a correct response, which is such a notable feature of complex learned behaviour. Hence, if we include 'type II' under 'conditioning', and then say that all learned behaviour is conditioning, we are merely uttering a tautology, since we are already using the term 'conditioning' to cover all kinds of learned behaviour.

Since all this has been understood for several decades, it might be thought that to labour the point now is unnecessary. The need arises from the persistence with which writers of textbooks, especially of physiology, cling to an account of conditional reflexes which was always misleading and is now wholly inexcusable. It is especially absurd to end a chapter on the brain with a summary description (as we have seen, a misleading one) of Pavlov's work, without mentioning any other kinds of behaviour. The appeal of the textbook version is presumably its simplicity and the corresponding impression it gives that the underlying processes are quite mechanical. A student of behaviour or of psychological medicine soon finds, however, that the actual facts of behaviour cannot be forced into so narrow a framework. Machines can be made which 'learn' in just the way described by the books; but certainly no mammals do so. Consequently, the student finds himself in a multi-dimensional labyrinth with no map. He is then almost inevitably forced to resort to unanalytical and unprofitable concepts such as 'instincts' and 'drives'; most of these will be without adequate definition in terms of observed behaviour or physiology. And so the chaos which has long surrounded the sciences of behaviour is preserved.

The importance of Pavlov's work, and that of his followers, lies in its method. It has made possible systematic, objective and rigorous studies of the learning process even in such large-brained animals as dogs. It also helped to reveal the complex of autonomic activations and other bodily processes which accompany the changes in the forebrain that directly produce the learned behaviour (cf. § 8.2.3). Despite this, the great mass of laboratory research on the 'laws' of learning have been on animals in trial-and-error situations: they have been done in mazes, problem boxes and the like. It is to the results of this work that we must now turn.

7.3 THE FACTORS OF LEARNING AND EXTINCTION

7.3.1 *The stimuli*

7.3.1.1 *The sensory modalities.* In animals with nervous systems all behaviour is mediated by links between sense organs and effectors. Learned behaviour depends on the creation of new links. We must now ask whether all sense organs can take part in the learning process: that is, whether *any* kind of 'cue' can be involved in a new association with a response. In most experiments on learning the cues are chosen for the experimenter's convenience. Visual and auditory stimuli (within certain ranges) are easily observed and presented, but not kinaesthetic or olfactory ones. Neglect of the preferences and limitations of experimenters can lead to error. An example of a sensory capacity which is easily ignored is the ability to respond to sounds of a pitch beyond the capacity of the human ear. Animals that can do this commonly also produce such sounds. These sounds may have two functions: first, they may act as social signals (§ 4.2.1.5); second, they may serve in echolocation. Small mammals are usually nocturnal, and so a non-visual method of locating objects at a distance is appropriate for them. The whole of this subject has been admirably reviewed by Griffin [115]. As for rats specifically, Anderson has shown that they produce 'ultrasonic' sounds [7]; and Riley & Rosenzweig have demonstrated that blinded laboratory rats can use echolocation while learning to run a maze [257].

It is not difficult to grasp imaginatively that other animals respond to sounds beyond our hearing; and echolocation by means of ordinary sounds is used by blind human beings. But comprehension of the 'olfactory world' of an animal such as the rat is far more difficult. We have seen that odours play a major part in mating and other social behaviour (§§ 4.2.1.4, 4.3.3.2) and that rats readily make use of odours in finding food (§ 3.2.1.1). Ritchie has further shown how important and even disconcerting such stimuli can be during experiments on learning: he was concerned with the question (discussed below) whether rats find their way about by (i) learning to make particular movements or (ii) responding to particular objects in their surroundings; eventually he discovered that the rats in his experiments were directing their movements by means of the odours and sounds from other rats caged in the same room [258].

In general, then, any exteroceptors can mediate in learning, and they do so according to their own range of sensitivity, which is often very different from ours. The central nervous system has, however, also internal sources of 'information'. These too can supply cues for learning. Blinded rats can discriminate inclined planes differing in slope by only ten degrees, and some can distinguish differences of only four degrees. Hence if rats are run in a maze in which the pathways have different inclinations they learn more rapidly than one of the same pattern in which all pathways are level. Rats can also be trained to select the turning which is a particular distance from the start, with no cue other than the distance. These are examples of the importance of kinaesthetic cues [294]. Further, we saw in § 2.2.3 that spontaneous alternation is usually based on visual differences between the alternative paths; but that in certain conditions it is possible to demonstrate response alternation, in which the cues are evidently proprioceptive. A number of studies, reviewed by Bindra [48, pages 181–2] have shown that interoceptive cues too can be the basis of learning: for instance, rats can be trained to avoid a shock by taking one path when they are hungry but another when they are thirsty, even though they receive neither food nor drink in either path.

7.3.1.2 *Direction-finding.* Any input to the central nervous system can, then, serve as a cue or conditional stimulus for a learned act. In any situation which offers a number of cues, affecting different senses, it is impossible to say *a priori* which are the ones that actually evoke the animal's responses. Arising from this difficulty, there has been a controversy on just what sort of cues a rat uses in learning a maze. Is it learning to make a particular series of movements? If so, the process may perhaps be described in stimulus-response, or SR, terms: that is, it consists of learning to make one particular movement to the stimulus situation at the starting point, another at the next point, and so on. This formulation has been put forward by a number of workers, notably C. L. Hull [156, 157]. It has long been known, however, that no such theory can account for all the behaviour shown in mazes or other situations requiring problem-solving. The early work on this subject was reviewed with elegant lucidity by Lashley [186] – lucidity which, unfortunately, few writers in this field contrive to imitate.

One kind of observation which can hardly be squared with any simple stimulus-response formulation concerns learning to respond to a pattern: the pattern may in principle be visual (a shape), auditory (a tune) or, presumably, tactile (a shape again), though we know little about the last apart from some work on *Octopus* [338]: in most experiments it is visual. We saw in § 6.4 that a rat may be trained to jump on to a platform towards a particular shape marked on a card, and that the precise shape can be varied to some extent without interfering with the response. This is an example of the very important phenomenon of stimulus generalization (to be discussed also in § 7.3.1.3). Familiar examples in ourselves are the recognition of letters printed in different type faces and of sentences spoken in different accents: each member of a series of events may be clearly different yet display a particular system of relationships, or pattern, in common. Another kind of example on the sensory side is the result of training an animal to make a brightness discrimination with one eye covered; on completion of the training, if the animal is tested with the 'trained' eye covered and the other free, the discrimination is still made.

On the motor side, correspondingly, we find response generalization. Rats trained to run a maze can make their way through it thereafter without necessarily repeating exactly the same movements: if the maze is filled with water, they will swim through it; if, after training, they

have operations that damage the cerebellum and so cause motor inco-
ördination, they will roll through it. Dodwell & Bessant give a still
better example. Rats were trained to swim through a water maze; some
had previously been run passively through the maze on a trolley and
these did better than controls which had not had this experience. After
the first three or four runs on the trolley the rats anticipated the next
turn by moving their heads; and, as the goal box was approached, they
moved to the front of the trolley or even jumped into the water. Here is
an example of learning without performing the activity (swimming)
which makes the learning evident to an observer [86].

Accordingly, some workers have supported the view that, in learning
a maze (for instance), a rat or other mammal is learning a set of relationships, just as a man does when he becomes able to carry a map in his
head. This view is especially associated with the name of E. C. Tolman;
and Tolman does indeed write of 'cognitive maps' when describing the
behaviour of rats [316]. This phrase illustrates very well how difficult
it is to describe some complex behaviour except in imagery based on
human experience. We have as yet no knowledge of the neural processes
which underlie the behaviour; no doubt, when we have, our language
will become more precise.

It was mentioned in § 2 that the ability to learn the way about a large
area is widespread in the animal kingdom. Many species of insects
display it, as do other invertebrates. The learning may be extraordinarily rapid [312]. The process is probably substantially different from
that in mammals. Maier & Schneirla [208] have described it in ants
(*Formica*). Schneirla writes:

> The insect builds its maze habit slowly and by stages: first a general
> habituation to the situation develops, then a process of very gradually
> mastering the local choice-points, and in a final stage an integration
> of the habit only through stereotyped interactions between [separate]
> habits. [270]

He also points out that, while rats readily learn to reverse a maze, ants
do not. Nor do ants develop disturbed ('neurotic') behaviour in confusing conditions.

The theory of 'cognitive maps' in mammals is sometimes referred to
as a theory of place-learning. It does not assume the learning of any
specific *movements*, but only the association of signs or cues with 'cogni-
tions' or 'significates'. The last two terms refer to hypothetical processes
in the brain – hypothetical, because they cannot at present be identified
with anything that has been directly observed.

Although it has often been implied that theories of place-learning and response learning are alternatives, this is not really the case. Both kinds of learning certainly occur. The evidence and controversies in this field have been fully reviewed by Osgood [239] and by Mowrer [230, 231]. We saw in § 7.3.1.1 that the cues for learning can be of all kinds, including proprioceptive. This conclusion may now be supported by describing two further groups of experiments which have a bearing on the direction-finding of rats in mazes. The first group includes those which deal with latent or exploratory learning, already

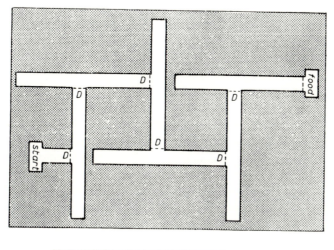

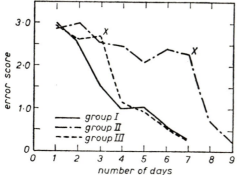

52. Exploratory (latent) learning. Above, plan of maze used by Blodgett. (D = doors which were closed behind the rat as it moved from start to goal.) Below, learning curves of three groups of rats. Those of group I were trained when hungry in the usual way, with food in the goal box on each run; their learning curve is typical. Rats in the other groups were not at first rewarded, but at point X reward was introduced: this led to a sharp reduction in errors, attributable to previous exploratory or latent learning. (After Blodgett [52].)

mentioned in § 2.4.1. The essential features of this behaviour were first made clear by Blodgett [52]. If rats are put in a maze, but are not rewarded on reaching the goal point, they explore the maze. If, later, they are run in the maze in the usual way, with a reward of food at the goal point, they learn the maze more quickly than controls which have had no unrewarded experience of the maze: they make fewer errors and their running time is shorter (figure 52). There has been controversy on this subject, reviewed elsewhere [21], but there has been plenty of confirmation of the reality of exploratory learning in the thirty years that have followed Blodgett's work.

The rapid learning of a maze which is the index of exploratory learning is an example of insight behaviour as defined in § 6.3. It shows that rats can behave as though they had learned their way about an area in the same way as a man exploring an unfamiliar town: they can use this experience when faced with a new situation which demands movement in a particular direction.

The facts of exploratory learning do not, however, tell us anything precise about the stimuli, external or internal, to which the rat in the maze is responding. The question of direction-finding by rats has, however, been studied further in a 'sunburst' apparatus of the kind shown in figure 53. In this situation Tolman and his colleagues found that rats chose most often the one path, out of eighteen possible paths, which led them directly towards the goal, although they had previously learned to reach the goal from a quite different quarter: the rats behaved as though they had a 'sense of direction'. This sort of experiment has been repeated by other workers, and it is now clear that success depends on the existence of plenty of cues outside the maze itself: as we saw in the previous subsection, rats can orientate themselves on all sorts of directional stimuli.

We may sum up this rather confusing subject by first repeating that rats can make use of any kind of stimulus, external or internal, in learning to find their way about. Secondly, we must agree that rats do sometimes behave like men remembering a map or plan or the layout of a system of relationships previously experienced. Thirdly, it follows, from this and much else, that learned behaviour cannot be adequately described in simple stimulus-response terms, any more than it can be 'explained' by conditional reflexes. Finally, it seems that if we wish to try to go beyond these conclusions, we are obliged to use terms which, if they mean anything, refer to processes in the brain; but at present this does not get us very far, because these processes have not yet been identified.

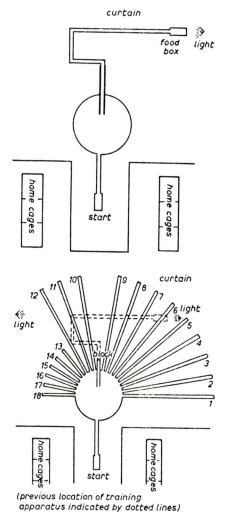

53. Test for 'sense of direction'. Above, plan of apparatus in which rats are trained to make their way to food by an indirect route. Below, apparatus in which they are subsequently given the opportunity to go directly towards the original goal (along path 6). (After Birch & Korn [51], based on the design of Tolman, Ritchie & Kalish.)

7.3.1.3 *Generalization and discrimination.* In most of what has been said so far it has been implied that when, for example, a rat is said to respond to the sight of an object by jumping, the object is always the same. In the preceding subsection, however, we saw, in the brief discussion of stimulus generalization, that this assumption is wrong. If animals responded only to external changes as narrowly defined as, say, the radio impulses which make a guided missile alter course, they would be hopelessly inefficient. In fact they respond to *populations* of stimuli; each population must have a mean value for each of its measurable attributes, and a certain degree of variation about the mean. The variation depends (i) on the objects or processes which provide the stimulation; (ii) on the distance and direction of the source of stimulation; (iii) on the animal itself.

As for the first of these, if the objects responded to are other animals they will certainly vary. The second source of variation is equally obvious: the angle or distance at which an object is seen, and the distance of the source of a sound, must influence the number and arrangement of the individual receptors (rods and cones, cells of the organ of Corti and so on) which are activated. The variation of this sort that can be tolerated may be considerable.

The third source of variation is the animal itself. Consider, with Osgood [239], an experiment in which a rat is standing on a grid, a tone of 1000 cycles per second is sounded and the grid is then electrified. The rat jumps. When training has made the rat jump at the sound alone, a tone of 900 c.p.s. is presented instead. A jump follows, but one less vigorous than usual. With further lowering of pitch the intensity of the response declines further; at 500 c.p.s. the rat merely tenses and quivers. A similar effect is produced if the pitch is raised. This is a simple, neatly quantitative, example of generalization.

The opposite process is discrimination. It is possible to train a mammal to respond specifically to a very narrow range of frequencies and to ignore tones outside that range. Such discrimination may be brought about experimentally in two ways. First, an animal may learn to respond to one stimulus, s_1, but not to another, s_2, measurably different from s_1, because response to s_1 is rewarded and response to s_2 is not. This is a case of conditional inhibition, such as was originally studied by Pavlov. Secondly, response to s_2 may be *punished*: s_2 will then come to evoke a response, namely, avoidance, which is incompatible with that produced by s_1.

An important feature of discrimination is illustrated by a study by Brunswik of the behaviour of rats in a T-maze. The rats were fed sometimes when they turned right, sometimes when they turned left, often not at all. The probability that a rat would turn right or left was found to be roughly proportional to the incidence of reward on one side or the other. Hence, up to a point, a rat's behaviour is as if it were based on simple calculations of probabilities. Of course, the whole of its past experience influences its selection of alternatives, and not only its experience of the immediate type of situation [60].

Generalization and discrimination are necessarily features of all complex learned behaviour. The precision with which an animal adapts its behaviour to complex situations must depend largely on these two processes. We saw in § 6.4 that generalization of shapes, that is, response to certain relationships shared by a number of different patterns, can be taken as an index of 'intelligence' or learning ability. What would ordinarily be called 'intelligence' obviously depends on discrimination too. If an animal is presented, over a period, with a great number of rather similar stimuli, of which only a minority are accompanied by reward; and if that minority have only one feature in common which is not present in others; then the animal is faced with a difficult problem of discrimination. The capacity to make such discriminations, within the scope permitted by the sense organs, is clearly one aspect of learning ability.

7.3.2 'Reward' and 'punishment'

7.3.2.1 *What is 'reward'* ? Generalization and discrimination ensure that an animal responds to an appropriate range of objects in a given class. We must now consider further what determines whether an animal will respond at all. We have seen that the response of an animal to a given situation depends on its internal state. If an experimenter wishes an animal to perform a particular act, he usually arranges that performing the act is followed by the satisfaction (if only partial) of a bodily need (figure 54). Typically, he creates a state of need by depriving the animal beforehand. We must now say something more precise and detailed about reward, or reinforcement, and its relationship to performance.

A positive reinforcer may be defined as a stimulus which increases the strength of a response. This definition obliges us to consider the meaning of 'response strength', or 'habit strength'. What do we mean exactly when we say that an animal has learnt to perform some act? Further, how do we represent the progress of learning against time or against exposures to the learning situation? Pavlov, in his most famous experiments, used as his criterion the amount of saliva secreted on successive presentations of the conditional stimulus. This is an example of the use of the *amplitude* of the response. A second measure is *latency*, that is, the interval between the application of a conditional stimulus (or situation) and completion of the act; this, or its reciprocal (the *speed*), is easily used in a maze experiment. A third criterion is *resistance to extinction* (§§ 6.2.1, 7.3.3). Unfortunately these criteria are not always highly correlated, as we shall see (§ 7.3.2.2). It follows that strictly one should always specify the index of response strength used, and not refer to response strength by itself.

The definition of positive reinforcement is so general that it covers a great variety of agencies. It is usual to distinguish two general classes of such reinforcers. *Primary reinforcers* are sometimes defined, not very

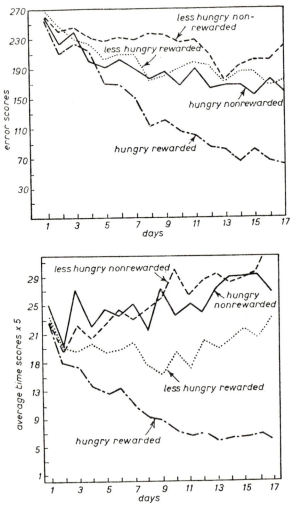

54. Effect of reward on learning. Time taken in a maze and number of errors made decline steeply with training only if the animals are both 'motivated' (hungry in this case) and appropriately rewarded (with food). (After Tolman & Honzik [318].)

satisfactorily, as those which are independent for their effect on association with any previous event: they provide the releasing stimuli for stereotyped acts. As we saw in § 5.3.2 the development of the ability to respond to a releasing stimulus may be complex, even when the ability is common to the whole species; hence it is likely that many reinforcers are called 'primary' when they do not in fact satisfy the definition. However, these reinforcers do form a fairly distinct class from *secondary reinforcers*, defined as those whose effects are acquired through association with another reinforcer.

We must now consider whether it is possible to make any valid general statements about the relationship of reinforcement to response strength. An obvious likelihood is that there would be a quantitative one, so that the greater the reinforcement the greater or quicker the response. Spence has summarized a series of experiments by Sheffield and his colleagues which support this notion. They recorded speed of running along a passage as a function of the sweetness, nutritional value and previous opportunity of drinking a 'reward solution' of saccharin (sweet) or of dextrose (nutritious but not sweet); they also measured the rate of consumption of the solution. There was a very precise relationship between response speed and amount of solution drunk, that is, the amount of consummatory response actually performed. In another study a similar relationship was found when the reward was the opportunity to copulate: speed was correlated with the vigour of the copulatory activity [298]. Young & Shuford studied the reinforcing effect of sucrose solutions of various concentrations; their criterion was again the speed with which rats approached the solution, but in their experiments the rats were sated, not hungry. They found that speed was a function both of concentration of sucrose and of length of time during which the rats were allowed to drink [340].

The results of these experiments are agreeable to common sense. Others, however, have shown that the relationship between reinforcement and response may be far from simple. O'Kelly & Heyer, for instance, studied maze-learning in rats deprived of water. Some had been without water for eleven hours; others, similarly deprived, had also received injections of sodium chloride to give a '36-hour thirst'. Maze-learning was the same in the two groups, even though the second group drank more water on each run. In this example the *duration* of the period of need evidently determined the strength of the learned habit, though it did not decide the intensity of the consummatory response [235].

Another kind of complication is found in the work of Crespi. He recorded the running speed of rats in a straight runway after they had had no food for 22 hours. The amount of food at the end of the runway (the reward) was varied. The speed of running evoked by 16 units of food was used as a base-line for comparison. If rats which had been trained on one or four units of food were then given 16 units, their running speed went *above* that of the rats trained throughout on 16 units. Correspondingly, rats trained on 256 or 64 units, and shifted to 16, came to run at a speed *below* that of the 16-unit group. These observations are most readily described in anthropomorphic terms, namely, elation and disappointment [77]. They imply, as do other experiments already described, that there are processes in rats' brains ('expectancies') which, on the basis of previous experience, anticipate the consequences of action.

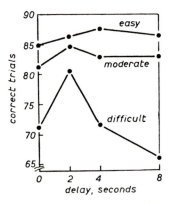

55. Effect of severity of discomfort in relation to difficulty of task to be learnt. Given a difficult brightness discrimination (bottom graph) rats performed best when held only two seconds under water before being allowed to swim on their own. (After Broadhurst [54].)

In Crespi's experiments the period of food deprivation was constant; hence it may be said that the level of motivation or 'drive' was the same for all rats (assuming that there was no significant variation between individuals). It might be supposed, especially in view of the work of O'Kelly & Heyer quoted above, that as the period of deprivation is increased so would response strength, at least until debility set in. In fact, the relation between duration of need and response is not so simple. Finan has shown that there is an optimum period of food deprivation for the learning of a bar-pressing habit. He used resistance to extinction as his criterion and during training deprived his rats of food for 1, 12, 24 and 48 hours respectively; the unrewarded trials to produce extinction of the response were all carried out after the rats had been deprived of food for twenty-four hours. The rats most resistant to extinction were those which had had twelve hours of deprivation during training [95].

We may then think of motivation as having an optimum level. What this level is has usually to be determined for each situation, but one generalization can be made about it: the optimum is lowered as the difficulty of the task to be learnt is raised. This, the Yerkes-Dodson law, first formulated in 1908, has been tested and discussed by Broadhurst. He gave rats a brightness discrimination which had to be made under water; the discrimination was made 'easy', 'medium' or 'difficult'. Four levels of motivation were contrived, by keeping the rats submerged in water for eight, four, two or no seconds before releasing them to swim through a T-maze. A correct choice enabled them to come to the surface and breathe. Figure 55 shows that for learning the difficult discrimination there was a marked optimum at two seconds of deprivation of air. Observations of this sort suggest that there are processes in rats (and other mammals) similar to those which accompany anxiety in man; and that (as in man) if these go beyond a certain intensity they impair learning ability. This is further discussed in § 8.5.

7.3.2.2 *Intermittent reward.* In the experiments described above, each correct response was rewarded. Skinner [290] has made a series of

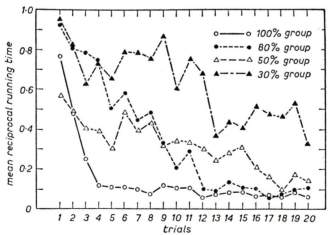

56. Extinction. The curves shows the rates at which a running response was lost by rats which had experienced different incidences of reward during training. Those rewarded on each trial had the highest extinction rate while those rewarded least often (on only 30 per cent of runs) lost the habit least quickly. (After Weinstock [328].)

studies of 'partial reinforcement'. He used, of course, the Skinner box. One method is to give the reward only once in a given *number* of correct responses (figure 56). This itself can be done in two ways. First, the intervals may be regular: for instance, every tenth response may be rewarded. In this case the response rate of a trained animal slows down just after reward but speeds up again just before the next reward is due: this is evidence, once again, of anticipation of the reward. Second, the intervals between rewards may be irregular and unpredictable: the response rate is then relatively steady and is little affected by reward when it comes.

Another way of applying partial reinforcement is to space out the rewards by time, regardless of response rate. This has a different kind of effect on behaviour. When the application of reward is determined by *number* of responses made, lowering the incidence of reward increases response rate: giving a food pellet once in a hundred responses produces a higher response rate than giving it once in ten. By contrast, if the reward is given once per hour, the response rate is lower than when the reward is given ten times per hour.

These observations are not very surprising, though they could conceivably have some practical applications. Of more interest is the observation, made by Mowrer & Jones [232] among others, that fifty per cent reinforcement can give as rapid learning as reinforcement of every correct response. The significance of this is obscure. Further, the effects of different sorts of partial reinforcement on response rate during training (and hence on learning) are not the same as their effects on extinction. We have just seen that the learning rate is faster when a high proportion of responses is rewarded. But, by contrast, the *extinction* rate is *higher* when the incidence of reward is higher. In other words, a learned habit is only slowly lost if the animal has been accustomed to being rewarded only seldom. This is a neat example of the need to specify just what feature of response strength one is using as a criterion in a given set of experiments or in drawing conclusions. It also illustrates the fact that 'learning', even of one of the conventional categories, covers a number of processes imperfectly correlated with each other.

7.3.2.3 *'Drive reduction'*. One may, then, make only tentative general statements about reinforcement and response strength; but it is possible to say something decisive about the notion, associated especially with C. L. Hull [158], that reinforcement depends on 'drive reduction'. This is sometimes called the 'law of effect', from its earlier version proposed by E. L. Thorndike (1874-1949). It is based in the first place on some obvious facts of physiology: the animal body is maintained, on the whole, in a steady state; for this to be achieved the external conditions such as temperature must remain within a certain range and certain substances must be available. Departure of any of these from a certain range of values sets up a state of need. The theory equates need state with 'drive state'; and reduction of drive state is held to be the invariable accompaniment of reinforcement. Although, perhaps, nobody now holds this view in just this form, it is still instructive to consider the evidence which bears on it. Even if simple theories of drive reduction are not tenable, they have the merit of encouraging the objective analysis of behaviour in terms of measurable functions, such as body weight or the level of substances in the blood.

We saw in § 2.2 that need, for instance for food, can increase activity: appetitive behaviour, as its name implies, is often a reflexion of need, and during this behaviour the animal learns how to satisfy the need, for instance by visiting a particular spot or by making some special movement. Once the need is satisfied the animal turns to another sort of activity or becomes quiescent. All this fits in with the assumption that need-reduction determines what acts an animal will learn to perform. But we also know (§ 2.2.5) that the opportunity to move around a relatively large area, or to watch moving objects, can itself act as a reinforcer. As far as *immediate* homeostatic needs are concerned, such acts are sheer waste of time. Admittedly, exploratory behaviour contributes indirectly to survival (§ 2.4.1); but at the time at which it is performed it merely depletes the animal's energy resources. It might be argued that there is an 'exploratory drive', and that exploration serves to reduce this drive. But such an argument becomes no more than a tautology: it implies that an act is performed only because it leads to drive reduction; and when we ask how we know that drive reduction has occurred the only answer is that it must have occurred – because the act was performed.

There are also many examples of reward value in activities which, by themselves, are of no biological use. An example is the performance of coitus without ejaculation. We saw in § 5.2.1.1 that male rats have to copulate a number of times before they succeed in ejaculating semen. Sheffield and his colleagues have shown that the opportunity to copulate without ejaculation acts as an effective reward in a learning situation (figure 57).

In the form in which it is sometimes stated, the theory of drive reduction is a misapplication of a biological principle. It is acceptable to say that everything one observes in an organism is a product of natural selection; and that it has, or has had, survival value. (It may be necessary to add that some features of organisms are *indirect* consequences of processes which have survival value.) But it does not at all follow that all behaviour directly and immediately contributes to homeostasis. If one considers mating and the care of young, this can at once be seen as a truism. But the point still holds even for (say) behaviour related to

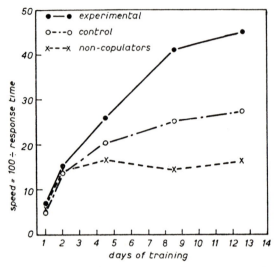

57. Reward value of coitus without ejaculation. The experimental animals found a female in the goal box of a maze; they performed coitus but were removed before they reached the stage of ejaculation. Control animals found a male. The bottom graph is the learning curve of animals which did not attempt coitus, whether they found a female or a male. (After Sheffield, Wulff & Backer [288].)

ingestion. We have seen, for instance in § 3.3.2, that rats display preferences for substances, such as saccharin, which are not related to their nutritional value. They can, sometimes, adjust their intake according to need, and generally do so as far as energy intake is concerned; but they are not infallible. This is to be expected. Natural selection operates so that the behaviour of an animal will, *in the conditions in which its ancestors have survived*, conduce to continued survival of the species; but the means by which this is achieved are diverse and may be very indirect. (This, of course, by no means precludes survival also in novel environments. Rats, for instance, were in some ways obviously pre-adapted for life in laboratories.) When an animal is subjected to quite bizarre conditions, its behaviour is likely to be bizarre too, and to fail to relate response to need. This is outstandingly the case in some celebrated experiments by Olds (figure 58). Micro-electrodes were implanted in the brains of rats and connected up so that each rat could, by pressing a lever, stimulate a portion of its brain. If the electrodes were in the right place the rats found this so enjoyable (to speak anthropomorphically) that they soon learned this and did a vast amount of self-stimulation [236].

The obvious conclusion is that, although an animal's behaviour reflects the action of natural selection, just how the behaviour is related to survival must be investigated in each particular case. It cannot be determined *a priori*. Accordingly, what constitutes an inducement to learn a task may have anything from a direct to a very tenuous relationship with survival.

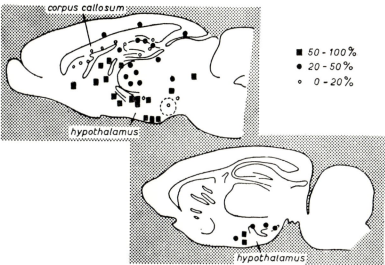

58. Sagittal sections of rat brain illustrating work on self-stimulation by rats. Electrodes implanted at points shown by black squares were highly 'rewarding'; black circles indicate moderate reward; and open circles neutral or negative effects. (After Olds [236].)

7.3.2.4 *'Avoidance conditioning'*. Despite this conclusion, it remains true that a good deal of learned behaviour can be related in a quite straightforward way to obvious bodily needs. The needs referred to in the preceding subsections have been mainly those which can be satisfied by *approaching* a source of stimulus. But there are others which demand flight. To put the point in another way: there are events outside the animal of which the *removal* increases response strength. Subjection of an animal to such events may be called 'punishment'. Sometimes however, these events are called 'negative' reinforcers, but here there is a difficulty. Literally, as Bindra [48] points out, a negative reinforcer might be supposed to be the antithesis of a positive reinforcer: that is, an event which *decreases* response strength. Olds, in the work quoted above, did indeed find regions of the brain whose stimulation *deters* the animal from repeating the act which caused the stimulation [236]. In what follows, despite its overtones, the term 'punishment' is used for events whose removal increases habit strength.

Observations on the effects of punishment are rather few. In some respects they parallel those on reward. Campbell, for instance, found that rats would learn to perform a task in order to reduce the amount of noise to which they were subjected and that the rate of learning was proportional to degree of noise reduction [64a]. Similarly, Campbell & Kraeling showed that rats which were obliged to run down an alley to escape a shock ran at a speed proportional to the shock decrement [65]. Other patterns are less predictable. There are examples, discussed by Bindra [48], of a tenacious resistance to extinction: an act which has led to punishment, such as electric shock, may evidently be profoundly inhibited even after one or a few experiences. In this, as with the reaction to noise, there are conspicuous similarities to our own species.

The resistance to extinction of some learned avoidance behaviour leads us to an important aspect of punishment. It has been suggested, by Mowrer [230] among others, that there are two distinct learning processes. On this view, the account, given above, of the effects of positive reinforcement deals with the results of one kind of internal process which leads to learned behaviour, but the effects of punishment are due to another kind. It is a fact familiar from human behaviour that if pain is experienced in a particular place, the place itself may come to arouse fear. Indeed, this is a usual consequence of punishment. While learning impelled by a need stems from a condition which is continually with the animal until the need is satisfied, learning to avoid an externally caused pain is inevitably related to the site at which the pain occurs or to other environmental conditions which accompany the pain. Miller, for instance, gave rats electric shocks which could be avoided if the animal went into an adjoining compartment of the cage. Each rat so trained was then put in the compartment in which shock had been experienced, but with the door to the other compartment closed; no shock was given; the door could be opened by turning a wheel or, later, by pressing a bar. These acts were readily learned by rats which had experienced shocks of moderate severity, but not so readily by control rats which had not been shocked. The trained rats, on being dropped into the first compartment for no-shock trials, urinated, defaecated and displayed signs of excessive tension; these objective signs are the grounds for speaking of 'fear' in these rats [217].

Experiments of this sort have been said to show that 'fear' is an 'acquirable drive', and that reduction of fear constitutes a reinforcement for the learning of new responses. This notion can be given, to some extent, a physiological meaning. At least since 1915, when W. B. Cannon published his *Bodily Changes in Pain, Hunger, Fear and Rage*, it has been known that injurious agencies activate the sympathetic nervous system and stimulate the secretion of adrenalin. If such a noxious stimulus regularly follows some previously neutral occurrence, such as being put in a box or hearing a sound, the visceral activation comes to be attached to the neutral occurrence: in other words the neutral stimulus acts as a conditional stimulus and the sympathetic activity becomes a conditional response. The components of the visceral response may include increased heart rate and respiratory rate, sweating and movements of the alimentary tract. These take place also in human beings when they are frightened or anxious, and so we have in them objective criteria of fear and anxiety. Mowrer's proposal is that when an animal learns to avoid a noxious stimulus, for instance by jumping from one compartment to another in a shuttle box, this is on the basis of an anticipatory visceral response: the latter immediately follows the conditional stimulus, and *ending the visceral response* (by departure from the alarming situation) becomes a secondary reward for flight.

We know well enough that certain stimuli cause immediate avoidance: cutting the skin is a simple example. Mowrer assumes that fear or anxiety, or rather the visceral responses just described, act in themselves as noxious stimuli which an animal will learn to avoid if it can. Consequently, it learns both to avoid anything which has become associated with this internal state, and also to take flight whenever it finds itself in a position in which the visceral response is aroused.

There is some indirect experimental evidence to support Mowrer's hypothesis. Wynne & Solomon studied the effects of shock on dogs: some of their animals had had their sympathetic nervous system removed; some had had drugs which reduced sympathetic activity; and some, the controls, had merely been subjected to shock. They found that both sympathectomy and drugs reduced the avoidance behaviour due to shock [335]. In another kind of experiment, Farber studied rats which had developed 'response fixations' as a result of being shocked in a T-maze. A response fixation is a compulsively performed movement, such as turning or jumping always to the right regardless of the actual circumstances. Farber proposed that these responses come to be made and fixed because they lead to 'anxiety reduction'; and that they do this because they take the animal away from external cues which induce anxiety on account of their association with shock. Farber also suggested that feeding would reduce anxiety; and he did indeed find that feeding his rats in the T-maze, in conditions in which response fixations were ordinarily developed, prevented the fixations [94].

We can now summarise Mowrer's 'two-factor' learning theory. As shown in the table below, he distinguishes two classes of stimuli:

	positive reward (hope)	*punishment (fear)*
response-correlated stimuli	response is augmented to give learned habit	response is inhibited
independent stimuli	approach	avoidance, flight

(i) those due to an action of the animal (response-correlated) such as pressing a lever – which gives rise to proprioceptive and other stimuli; (ii) those independent of the animal, such as a noise or attack by a predator. The 'stimuli' in this context include internal processes, even some which go on in the brain. If a stimulus of the first category is followed by reward then the act that led to it tends to be repeated and so a learned habit is produced; while a stimulus of the second sort, also rewarded, induces approach to the source of the stimulus. In either case, in Mowrer's terminology, the stimulus gives rise to 'hope', that is, anticipation of reward. By contrast, punishment of a response-correlated stimulus causes inhibition of the response; while injury associated with an independent stimulus leads to flight. In both the last two instances fear is aroused (that is, anticipation of pain).

Two features of Mowrer's account require comment here. One is the anthropomorphic terminology. This reflects a feature of mammalian ethology which must strike any detached observer: it is the way in which objective studies of rats and other mammals reveal, after the most exacting experimental and theoretical struggles, behaviour which is already perfectly familiar in ourselves. Naturally, we already have a colloquial (and 'subjective') set of terms in which to describe the behaviour – for instance, fear, anxiety, hope, elation and disappointment.

This does not, however, signify that the research so brought forth has been useless: only the patient accumulation of the facts, however bathetic some may turn out to be, can give us a foundation for theory and so a science of behaviour.

The second feature of Mowrer's system is the emphasis on the rôle of anticipation in learning. This is reflected in the use of terms such as 'hope' and 'fear'. We may now consider anticipation as a separate topic.

7.3.2.5 *'Expectancy'*. The anticipatory character of learned behaviour was first studied rigorously in the CR. Other examples too have already been given. Systems in which anticipation is emphasized are sometimes referred to as 'expectancy theories'. They may be explained first from the standpoint of common sense. If a hungry man finds food in a particular place, he will be more likely to revisit that place when he is again hungry; when he does repeat his journey he is said to be expecting food. We may assume that a specific pattern of change (presumably of facilitation and inhibition) in the central nervous system is set up by the combined effects of the internal state (hunger) and of the external situation associated with getting food. Actually finding the food 'confirms the expectancy' (subjective) or 'abolishes the neural pattern' (objective), and so strengthens the habit of visiting that place when hungry. Hence it might be said that anticipation during appetitive behaviour accounts for the effects of a reward in stopping the behaviour [312], since the reward puts an end to the anticipatory central nervous processes which impel the behaviour.

These general statements by themselves do hardly more than suggest a comparison of the behaviour of mammals generally with that of man. They are of little use unless they can be more specifically related to actual behaviour. The assumption is that animals such as rats, as well as men, have what Hebb has called an 'anticipatory central action' [133] during the performance of a learned habit. We have already seen that the behaviour of wild rats may be profoundly disturbed by an encounter with an unfamiliar object in a familiar place. This might be regarded as a disrupting effect on central nervous function of something that departs from an 'expectancy'. Further evidence is found in the phenomenon of motor equivalence. If a rat is trained to press a lever it does not always do so in the same way: just as a man may press a switch with either hand or even his elbow, a rat may use either forelimb or even its teeth; considerable practice is required to achieve a highly constant pattern of learned movements – far more than is needed to establish a uniform degree of success in actually getting the lever

depressed. Thus the process described in § 6.2.3 as a gradual elimination of all but the most effective movements in the appetitive phase is
usually not complete.

The importance of the anticipatory process has been strikingly
illustrated in an observation by Prokasy. He used a maze with a single
choice point in which the two goal boxes were both out of sight of the
rat even after it had chosen the left or right path. There was an even
chance that a given goal box contained food; but on *one* side the rat
could tell in advance of reaching the goal whether it would get food,
since the passage was painted white when food was there, black when it
was not (or vice versa). Figure 59 shows that after fifteen days of training
the rats developed a preference for the arm of the maze in which they
received advance information, even though they got no more food on
that side than on the other. Hence, as Mowrer [231, page 188] points

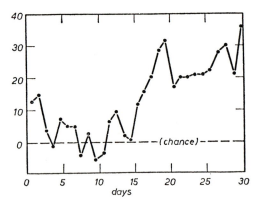

59. Expectancy. Rats come to prefer a pathway which gives them early information on whether there is reward or not, even though selecting that pathway does
not lead to reward more often than the alternative. In the graph, points above the
chance level show the degree of preference. See text. (After Prokasy [245].)

out, access to 'information' was rewarding. The long delay before the
habit developed is, perhaps, something that needs further study.

These facts may be linked with insight behaviour (§ 6.3) and also
with the notion of 'cognitive maps' put forward by Tolman to describe
direction-finding in rats (§ 7.3.1.2). In both there is, to quote Hebb
again, 'an association between a present state of affairs and one that has
followed that state of affairs (or similar ones) *in the past.*' In both we
may see the performance of a system of movements, in a novel situation,
to bring an animal to a specific goal. Once again we observe the ability
of animals, other than man, to make use of previous piecemeal learning
to synthesize an effective action from a number of previously separate

movements. This applies to the effects both of reward and of punishment. In each case the anticipatory process in the brain is a representation or model of the actuality which the animal has experienced in the past.

The ability to produce novel and appropriate patterns of action is one of the most difficult features of behaviour to account for neurologically; and it is the more baffling since, as we shall see in § 7.4, it probably depends largely on a special kind of learning process in early life.

7.3.3 *Extinction and recovery*

It has already been mentioned, in § 6.2.2, that complex learning always involves inhibition as well as the development or fixing of responses. We also saw that one form of inhibition occurs when a CR is repeatedly evoked without reinforcement. This important phenomenon, extinction (internal inhibition), has been studied also in trial-and-error situations. Osgood [239, pages 336 e.s.] gives the example of a hungry rat in a Skinner box. The response of bar-pressing is established by rewarding it with, say, a food pellet. Then the supply of pellets is cut off. The rat, after several unrewarded trials, resorts to appetitive behaviour such as sniffing around, biting the bar and so on; the bar is pressed only at irregular intervals. Eventually a substantial time passes during which the rat does not press the bar once, and extinction is said to have occurred. The criterion of extinction (that is, the length of time during which the act is not performed) is arbitrary.

After, say, twenty-four hours, the rat is again put in the box, and then displays spontaneous recovery, like a dog in a CR experiment. The inhibitory process which produces extinction has declined, and response strength is nearly, but not quite, what it was before extinction began. A curious feature of extinction is that it is more readily achieved the more energy is involved in performing the act to be extinguished: the more effort the animal has to make during the response, the easier it is to induce it to give it up.

Since spontaneous recovery is never complete the animal is evidently *learning not to respond* during extinction: as we saw in § 6.2.2, extinction is not an undoing of a learning process in the central nervous system. (To what extent such undoing ever occurs is a problem in itself.) The neural basis of the two processes, learning to act and learning not to act is, however, quite obscure. Certain drugs have nearly opposite effects on them: depressants such as sodium bromide, NaBr, produce slower positive learning but quicker extinction; stimulants, such as caffeine and amphetamine, promote positive learning but damp down the rate of extinction. These facts perhaps strengthen the belief that positive learning and extinction are different processes physiologically; but we have yet to learn how these drugs act on the brain. In any case, the term 'extinction' alone probably covers a diversity of processes: it comes under the heading of 'habituation', as that term is defined in § 6.2.2; and we saw there that habituation includes a number of different phenomena.

It was shown in § 7.3.1.3 that the choice by an animal both of objects to respond to and of responses to make can be thought of in terms of probabilities. Learned behaviour depends in a quantitative way on previous experience. This is true for past experience both of reinforcement and of absence of reinforcement. The preceding facts are, like other aspects of learned behaviour, enigmatic as far as physiology is concerned; but they represent ways in which an animal such as a rat is continually able to adjust its behaviour in accordance with the changing chances of its environment.

7.4 THE ONTOGENY OF LEARNING

In experiments on learning, changes of behaviour are usually observed over periods which are short compared to the animals' life span. As a rule a quite specific situation is repeatedly applied; it is the progressive, or sometimes sudden, change in response to this situation that we colloquially call 'learning'. We must now turn to a different time scale and to the effects of non-specific stimulation. We already know that stereotyped activities such as mating and maternal behaviour often depend for normal development on special conditions in early life; these provide examples of the effects of early experience on fixed action patterns. We must now consider the influence of conditions in early life on later learning ability [173].

This is a development of the notion that much learned behaviour depends on combining the effects of separate previous experiences (§ 6.3). A simple kind of this insight behaviour is where two portions of a maze have been independently learnt; but there is also the more generalized transfer of training displayed when learning one task increases the ability to learn another, different one. This kind of effect is not confined to topographical learning. It is the general phenomenon of 'learning set'. Harlow [123] and his colleagues have made a study of it in primates. In rats it has been demonstrated by Koronakos & Arnold. Rats were fasted and put in an apparatus in which there were five doors each marked with a pattern; four doors had the same pattern and the fifth a different one: in one instance four doors were marked with a rectangle and one with a cross. The door with the odd pattern was the only one unlocked; once through it the rat was required to make another, identical discrimination to get to food. Eight different 'oddity problems' of this sort were presented; each had to be mastered to a specific criterion before the next. Out of twenty rats, five clearly improved in the ability to learn the tasks as the experiments proceeded: in them, therefore, there was transfer of training or the formation of 'learning set' [176].

This kind of learning can occur in adults. A less specific form of 'learning set' has been called 'learning to learn', or 'deutero-learning'. Its importance is greater than is implied merely by results such as those just described. This has been made clear by rearing young rats in diverse environments and observing the effects on their ability to learn as adults.

Hebb [133, pages 109 e.s., 228] has made the major contribution to knowledge of early learning. His views apply, possibly, only to mammals, since they are linked with a distinctive feature of the mammalian brain, namely, the vast mass of connecting neurons which intervenes between the sensory input and the motor output. It would, however, be interesting to know more about other groups, such as birds, fishes and cephalopods. Hebb has suggested that much of the learning that goes on in very early life is different in kind from that which, later on, enables an animal to adapt itself quickly to new circumstances: he describes the early learning as both inefficient and slow, and yet responsible for the efficiency of later learning; at this stage no insight behaviour is, or can be, displayed.

An early study by Hebb, already quoted in § 6.4, arose from the famous work of Senden on the results of operations on human beings (*Homo*) blind from birth. Such people, on having sight conferred on them, are unable to distinguish objects (however familiar by touch) when they try to do so by sight; they have very slowly to learn to distinguish even, say, a square from a triangle. Hebb showed that rats reared in the dark took longer than controls to learn a simple visual discrimination. However, the effect of being reared in the dark soon wore off: after only one hour the rats' behaviour was normal [130]. This speed of recovery is in marked contrast with what happens in man, or even in the chimpanzee (*Pan*) [133]: these, with their far greater mass of brain tissue, have a much longer period of helplessness in infancy and take correspondingly longer to 'learn to see'.

The work on vision is a special case: it suggests that deprivation of the opportunity to use the eyes (and the corresponding parts of the nervous system) interferes with visual learning. More recent studies have gone further. Hebb himself, Bingham & Griffiths [49] and Forgays & Forgays [97] have all shown that learning ability in rats is much influenced by the diversity of their environment during their first few weeks. If they are allowed to experience a variety of objects they are more 'intelligent' in later life: that is, they can learn tasks in novel conditions more readily than controls kept in a restricted environment. This statement could justly be made in a different form: the 'natural' environments in which wild rats live provide plenty of diversity; one might therefore say that the experimental or peculiar conditions are those in which rats are deprived, by confinement in small cages, of the opportunity to 'broaden their minds'.

Forgus has provided more detail on the effects of early experience and on just what sort of stimuli improve learning ability (figure 60).

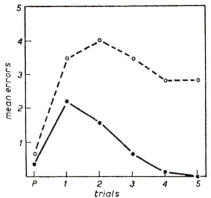

60. Effect of early experience. Rats of group I (lower graph) had had much visual and motor experience in early life; those of group 2 (upper graph), only visual experience. The graphs show the mean errors during successive trials in a complex maze in the dark. P stands for preliminary trials. (After Forgus [99].)

Improved 'intelligence' accrued whether the more complex environment could be entered or merely seen; this finding extends the observations of Hebb, quoted above, on the rôle of visual stimuli in young rats. But motor experience too had a favourable effect. One group of hooded rats was given much early visual and motor experience in a large and complex environment; a second group had similar visual experience but much less opportunity to move about. When the rats were tested, as adults, in a maze, the second group did the better of the two if the lights were kept on; but the first group did better in darkness, that is, in conditions in which learning had evidently to be based on kinaesthesia.

A further item in this work concerns the importance of the precise age at which diversity is experienced. Rats were given the opportunity to see a number of different shapes and patterns from sixteen days (when the eyes opened) to forty-one days; another group were similarly exposed from forty-one to sixty-six days. Both groups were later tested in conditions demanding discrimination or generalization of shapes, and both did better than controls which had had no early 'training'; but the first group, with the earlier experience, did better than the second [98-101].

The work on rats is paralleled by a smaller number of studies on other species, reviewed elsewhere [21, 46]. Dogs (*Canis*) and chimpanzees (*Pan*) have been used in formal experiments. It is unfortunate that we have little precise information on our own species, however much we may guess about the behaviour of children. Despite the few species so far studied, we can be sure that for mammals in general early deutero-learning is of prime importance.

7.5 REVIEW OF 'LEARNING'

Individual adaptability to changing conditions is displayed by all animals. Change of behaviour in response to stimulation is an aspect of this adaptability which can be observed in most animals. Such learned behaviour is at its most complex in mammals, but within this class (even if we omit our own species) there is much variation in learning ability. Rats are, among mammals, not especially notable for their intelligence; but they have proved to be excellent material for the analysis of learned behaviour in the laboratory; and wild rats display a repertoire of actions and abilities characteristic of small mammals.

At the behavioural level, as distinct from the physiological, the difficulties of even describing learned behaviour are best exemplified by unconfined wild rats. Cottam has described a large colony living in burrows near ponds and streams heavily stocked with fish. When food was thrown in the water, young fish concentrated in vast numbers. The rats learned to assemble at the same time and to eat scraps of food and catch the fish. The rats swam 'well and rapidly' [74]. This is a naturalist's example of complex learned behaviour resembling that given at the beginning of § 6. There is nothing stereotyped about the total pattern, though some of the component parts of the behaviour such as simple reflexes, must be stereotyped; other components, if the performance were analysed in detail, would certainly include CRs, habituation to a variety of stimuli and some compound of trial-and-error and insight behaviour.

These are the conventional categories of learned behaviour. The factors which influence learning have been revealed mainly in simple trial-and-error situations in the laboratory. They have shown that sensory information of any kind, from external or internal sources, can be used in learning; and that inhibition, extinction and reinforcement, first systematically demonstrated by Pavlov in CRs, are displayed also in trial-and-error behaviour. The rôle of the internal state, or motivation, is evident in both kinds of situation.

Trial-and-error, however, gives much more information about the learning process than the Pavlovian method: it allows the animal to display appetitive behaviour and so to acquire a new pattern of activity. The new pattern is as a rule established only when performing it is accompanied or followed by the attainment of a consummatory act or state; alternative patterns not so rewarded are given up.

This process involves *selection* of the stimuli to respond to; other quite different stimuli come to have no effect (habituation). Selection of some from among a number of similar stimuli is the important process of discrimination. But the opposite process also occurs: a response may generalize to stimuli which resemble the original one but depart from it quite substantially. The likelihood that an animal will respond in a specific way to a specified stimulus pattern may depend quantitatively on the probability of reinforcement, as determined by past experience. This process of selection is entirely different from the traditional concepts of learning by association; and the notion that learning in general can be described in terms of Pavlovian conditional reflexes is quite erroneous. This is still more obvious for insight behaviour, in which two or more previous experiences are used in the performance of a novel sequence of acts. This sort of behaviour seems to involve an 'anticipatory central action' or 'expectancy' concerning the consequences of further action.

One kind of incentive to learn is avoiding pain externally caused. This differs from ending an internal discomfort. It is often achieved by flight, but can lead to learning a more specific response. Inflicting pain may lead to generalization of the fear-and-flight response to the whole situation. (If flight is prevented, severe disturbance of behaviour may follow. See § 8.2.3.)

Learning depends to a substantial extent on reinforcement; and, since some of the most effective reinforcers are those which satisfy bodily (homeostatic) needs, it has been proposed that learning takes place only if there is reinforcement. It is, however, certain that internal changes which bring about adaptive change in behaviour can take place during general exploratory movements, without reinforcement: this becomes evident when exploratory learning is observed. Further, the variety of incentives which do promote learning do not all directly contribute to survival, either of the individual or of the species. Although natural selection may be assumed to ensure that behaviour tends to promote survival within a certain range of conditions, it does not follow that it will do so infallibly in all conditions.

The importance of learning processes which, like those that occur during exploration, are not evident at the time they occur, has been further shown in studies of the effects of early experience: in infancy a mammal undergoes a form of learning, seemingly both slow and inefficient, which enables it to learn rapidly and efficiently in later life; this deutero-learning is aided by subjecting the young mammal to a spacious and diverse environment.

Two formidable sets of problems remain to be discussed. One concerns what used to be called the 'affective' aspect of behaviour and is dealt with in the next chapter; and in § 9 the question of the neural basis of behaviour is examined.

8

The Analysis of 'Drive'

... we may conclude that the philosophy of our subject has well deserved the attention which it has already received from several excellent observers, and that it deserves still further attention, especially from any able physiologist.

CHARLES DARWIN
(*On the Expression of the Emotions in Man and Animals*)

8.1 THE CONCEPT OF 'DRIVE'

8.1.1 *Facts and definitions*

Every act performed by an animal, and every failure to respond to a stimulus, raises the question of what makes an animal either responsive or immobile. The term 'motivation' has already been used to name all the internal processes which make an animal active rather than inactive, and active in a particular way. This definition is, however, so general that it hardly delimits a manageable subject for study; it does not convey why a concept of motivation is needed. The need arises in particular from the *fluctuations* which occur in an animal's responsiveness – fluctuations often independent of changes in the animal's surroundings. A female rat meeting a male on one occasion allows or encourages him to mount her; on another she ignores him until he makes advances, when she kicks him off. A fragment of food is eaten on one occasion, but a similar fragment, at another time, is used as material for nest-building. This sort of change of behaviour is not a consequence of learning: it is independent of any special experience, and is most clearly exemplified in variations in the readiness with which certain fixed action patterns can be elicited: there are, for example, the alterations of behaviour which take place during the oestrous cycle and during pregnancy and lactation.

Since our descriptions of events depend on comparisons, we inevitably try to relate the movements of animals to those of systems that we understand better, such as machines; but our ignorance of the internal processes which determine behaviour makes such comparisons difficult to use. In an engine the oxidation of coal or oil releases energy which is transformed into motion according to the laws of physics; but we have no corresponding account of the actions of the several thousand million nerve cells in the brain of a mammal. There is a temptation to fill this gap by giving names to hypothetical entities which can be invoked to 'explain' behaviour. In § 5.4 we saw that this sort of situation has arisen from some uses of the word 'instinct'. We are now obliged to examine in the same way the behaviour which has led to the use of the word 'drive'; and we must also consider the meanings of 'drive' which can be inferred from writings in which it is used.

Following Hinde [150] we may list six classes of *facts* which are commonly described or 'explained' in terms of drives. (i) First we have the spontaneity of behaviour already discussed: a sleeping animal may (for example) wake up, groom itself and then move off towards a source of food or water although there has been no change in its surroundings. (ii) Hardly to be distinguished from this is change in response to a constant stimulus, such as the alterations in attitude to food or a potential mate mentioned above. (iii) The consequent 'directiveness' or 'goal-seeking' tendency of behaviour is most clearly expressed in the persistence of activity until a particular state is attained; but directiveness is also obvious when, as a result of previous learning, a particular path is chosen from a number of alternatives. (iv) A further kind of persistence of activity is often displayed on the removal of a stimulus which has aroused a particular behaviour pattern: the animal does not become quiescent, but moves around until it once again finds the lost object or another similar one; for instance, if a male approaches a female in oestrus and she moves away out of sight, he looks for her. Sometimes an alternative activity is performed instead (§ 8.2.2). (v) Some fixed action patterns form groups, of which all the members are activated at the same time: an obvious example is the group of activities displayed by a female with young; another kind of example is the fact that, in a male, aggressiveness and sexual potency develop together. The fact of simultaneous activation suggests that the different behaviour patterns have a common source. (vi) Finally, there are the effects of 'drive' on learning, discussed in the preceding chapter (§ 7.3.2).

Some of these phenomena suggest, by analogy with machines, that they are products of some sort of stored or dammed up energy which impels the animal to act until the energy is expended or neutralized. This notion raises many questions, of which some come partly from attempts to invoke a unitary principle to explain the whole diversity of facts. The term 'drive' has, through this striving for an orderly interpretation, often been used as though it referred to internal processes or agencies responsible for the spontaneity of behaviour, its persistence and the rest, and not to the behaviour itself. It is now well understood (though still worth repeating) that introducing a term to name hypothetical processes is in no sense an explanation of anything. Such a term can, however, be used to state a hypothesis. In the present case the hypothesis might be that if we knew more about the central nervous system we should find a single class of processes responsible for the phenomena listed above. These processes might then be called 'drive' or 'drives'.

This proposal is open to at least two criticisms. First, it is vague. Second, it leads readily to ambiguity in the use of the word 'drive'. For example, as Hinde [148] has shown, the contexts in which 'drive' occurs imply at least two meanings for the term: (i) it may refer to the state of internal organs, usually no doubt the central nervous system, but (ii) it may also refer to internal agencies (less often, external ones) which influence the state of the central nervous system.

The first meaning is usually implied in the common use of expressions such as 'hunger drive' and 'sex drive': here the internal processes referred to are those especially responsible for particular kinds of behaviour. A similar meaning is found also in phrases such as 'general drive state'. While the behavioural index of 'hunger drive' is usually the readiness with which an animal will perform some act which brings it food, 'general drive' concerns total activity. The status of the term 'drive' is in fact similar to that of 'instinct' (§ 5.4).

However, some ways of using the term are free from ambiguity. The phrase 'hunger drive', in a suitable context, is a convenient one. The context must be one in which the criterion of hunger drive, that is, the method of measurement, is completely clear (§ 3.4.3.4). It would therefore be an excess of pedantry to refuse to employ this expression. Similarly a common phrase such as 'reduction of drive' is convenient: it signifies a reduction in the readiness with which some activity can be evoked. Again, the behaviour in question must be specified exactly. However, some ethologists prefer to use the term 'specific action potentiality' (SAP) instead of drive. But the important thing is to define whatever terms are used, and to hold firmly to the definitions given.

8.1.2 Models and hypotheses

Whether the term 'drive' is ambiguous or not, we still have to consider the more important question: can the various phenomena to which it is usually applied be given any kind of unitary explanation? More than a century ago Herbert Spencer wrote of 'nervous energy' and 'nerve force' – phrases taken up by Darwin in *On the Expression of the Emotions*. Darwin wrote that (in Spencer's words) it was an

> 'unquestionable truth that, at any moment, the existing quantity of unliberated nerve-force, which in an inscrutable way produces in us the state we call feeling, *must* expend itself in some direction – *must* generate an equivalent manifestation of force somewhere'; so that, when the cerebro-spinal system is highly excited and nerve force is liberated in excess, it may be expended in intense sensations, active thought, violent movements, or increased activity of the glands. [82]

This passage has an antique sound, and is not in terms that a physiologist would find it convenient to use today; nevertheless, it implies that some behaviour can be explained by reference to quantitative changes in the distribution of some sort of energy. The same notion is the basis of a number of mechanical analogues that have been used in the description of behaviour. These analogues have been applied especially to the build-up of 'drive' (that is, readiness to act) that sometimes occurs when an animal has no opportunity to perform a stereotyped activity such as eating or coitus. Hydraulic models have been proposed by various authors: in one form, a reservoir fills gradually until a threshold is reached, and the machine or animal will then perform the act in question provided an external releasing stimulus is present; to parallel the facts of behaviour, a good many complexities of valves and keys are needed.

There has been much discussion about the significance of these models [152]. On the whole, the tendency has been to regard them as metaphors with the only function (if they have a useful function at all) of clarifying descriptions of actual behaviour. Certainly, the 'build-up of drive' is often a very impressive phenomenon. A striking example is the behaviour of previously celibate adult male wild rats when adult females are introduced into their cage: as one might expect, the vigour and persistence with which they approach the females far exceeds what is usually observed [26].

Many other phenomena in biology have this quality. Any process which goes on until a threshold is reached and a new effect is evoked may be described in the same way. An example is the build-up of central excitatory state in a neuron until the cell discharges and an impulse is produced in the axon. The reservoir model, indeed, does include one feature which is accurately paralleled by some events in the body: this is the accumulation of a substance (water in the model). Haldane has pointed out that the readiness of a centre in the brain to produce nerve impulses may well depend on the accumulation of substances; the accumulation of carbon dioxide in the blood has just this effect on the respiratory centre in the brain stem. Further forward in the brain is the hypothalamus, with its complex of centres 'controlling' ingestion and other behaviour and bodily processes. These may be sites at which accumulated substances are active [119, 120]. Reference has already been made, in § 3.4.5, to the effects of injecting minute quantities of substances such as sodium chloride into the hypothalamus; these show a way towards testing this kind of hypothesis.

8.1.3 *The plurality of 'drive'*

Hypotheses about the accumulation of substances ignore at least one major kind of fact: they try to account for both the initiation of behaviour and its ending, when in fact activity often ceases, not as a result of the exhaustion of a substance or source of energy, but because a particular input reaches the central nervous system. As we saw in § 3.4, the act of swallowing food can have a satiating effect even if the food never reaches the stomach; and the mere inflation of a balloon in the stomach may similarly reduce readiness to eat. Similarly, in mating, orgasm is followed by the ending of the activity as a result of afferent impulses which evidently inhibit central nervous activity. These are examples of peripheral agencies which have a 'switching off' effect. We are therefore in no position to talk about the impulse to act, or to become immobile, as though it were due to any single process. To talk about the 'hunger drive', for example, as if there were a single 'mechanism' responsible for seeking food, beginning to eat it, continuing to do so and finally stopping, is misleading.

In addition to the facts given in § 3.4 there is an instructive example in the work of Kennedy [169] and of Teitelbaum. The latter made lesions in or near the ventro-medial nuclei of the hypothalamus and so brought about increased food intake and consequent obesity. He observed, not only the amount of eating done, but also the amount of locomotor activity related to food-getting; and he found that 'hyperphagic animals . . . show a *lower*-than-normal drive to obtain food' [emphasis added]. This (seemingly anomalous) conclusion applied both to 'random' activity and to the amount of bar-pressing the rats were willing to do in order to get food [307]. Here, then, the 'drive' to carry out appetitive movements is independent of, or at least not positively correlated with, the 'drive' to eat.

Other aspects of the plurality of drive have already been mentioned in other contexts. We saw in § 7.3.2 that an animal can learn to perform an act which is not only useless but even harmful: behaviour is not inevitably linked to biological advantage. Similarly, drive reduction does not occur *only* when some direct advantage is achieved. This is the case with exploratory behaviour: a rat's tendency to explore declines as exploration proceeds, even though nothing useful has been found. There are also many instances of incomplete performance of a consummatory act being followed by drive reduction. Sheffield and his colleagues, in work already quoted (figure 57), allowed sexually inexperienced male laboratory rats to copulate with females but removed them before they achieved ejaculation. This acted as a reward in a learning situation even though it is also followed by disturbed behaviour.

Yet a further group of examples is found in the study of behaviour which seems to have gone wrong. It is to this subject that we must now turn.

8.2 'ABNORMAL' AND 'SUBSTITUTIVE' BEHAVIOUR

8.2.1 *The meanings of 'abnormal'*

There is a rather heterogeneous group of activities all of which may be called 'abnormal'. This term has many meanings, including 'atypical'; 'departing from the mean'; 'unnatural'; 'diseased'; 'biologically disadvantageous'. Here we are concerned with activities all of which *seem* (from the point of view of survival) to be inappropriate to the needs of the animal or species; we cannot, however, say that they are certainly useless or harmful, since we do not know all their physiological effects. If we see a dog panting with its mouth open we may infer that it has just been exerting itself vigorously; but it might be that it is cooling itself; and a third possibility is that it had just been subjected to a stimulus that arouses autonomic and allied activity ('anxiety' as described in § 7.3.2.3). If the panting is due to the last cause then we may regard it as a by-product of mobilizing the body for an emergency (as Cannon described); but we cannot be certain that this is its only significance. We are familiar, from human behaviour, with acts which, ostensibly futile, are of value to the performer because (we say) they release tension. We shall see that this sort of thing is possibly not confined to man.

Two main categories of behaviour fall to be described here. The first, sometimes called displacement behaviour, has been mainly studied by zoologists – especially in birds and fishes; the second, to which Pavlov's term 'experimental neurosis' may be applied, has usually been the preserve of psychologists.

8.2.2 *Substitutive behaviour*

The original notion of displacement behaviour has been reviewed by Tinbergen [314]. An animal, let us say, is performing some stereotyped act, such as mating, but is interrupted by the flight of the partner; on this, the animal turns abruptly to another activity, such as nest-building, eating, grooming or preening. The new behaviour seems useless and is often performed in an atypical way: it may be hurried or incomplete. There is an obvious, if superficial, similarity to the behaviour of some human beings when frustrated. This sort of thing may be seen, not only when animals are interfered with, but even when they spontaneously stop a fixed behaviour pattern (plate 12). Wild rats, during a series of copulations, may break off for a few seconds, run to food, eat hurriedly and then resume [19]. A herring gull (*Larus argentatus*) plucks grass at the edge of its territory in circumstances in which it might be expected either to attack or to flee; this is either a nest-building act (but no nest is in fact built) or a redirected attack on the opponent's feathers. There are plenty of such instances.

Unfortunately, rats, and mammals generally, are a poor source of examples, since detailed quantitative studies have not been made on them. There are activities which certainly would be called displacement behaviour as it was originally described: they include casual digging or scuffling of loose material such as grain [18, 70]; grooming and scratching; and perhaps the hoarding of useless objects, as described in § 3.2.4. In rats and other rodents a perfunctory brushing of the vibrissae is usually, perhaps always, seen when exploration of an unfamiliar area is temporarily broken off; it also occurs with great regularity when coitus or fighting are interrupted.

An important question is whether these and similar activities can usefully be regarded as belonging to a single class; to settle this we need to examine the internal causes of the behaviour, as well as the overt activity. Displacement eating in wild rats or displacement grass-pulling in gulls cannot be explained in terms of any known nervous processes; but other displacement behaviour can be related to autonomic activity. We have already seen that stimuli which provoke flight (fear) also cause activation of the sympathetic nervous system; such activation is followed by characteristic forms of behaviour. An example may be taken from the work of Andrew on buntings (*Emberiza*): when frightened these birds perform responses which usually help to cool them down when they are too hot, namely, gaping, feather sleeking and raising the wings; this behaviour is not due to exertion, but is evidently a by-product of the autonomic activation which accompanies fear. Sometimes warming behaviour occurs instead of cooling [9]. Such behaviour is comparable to paling or flushing and sweating in man when these have no thermoregulatory function. A parallel example in laboratory rats is given by Thompson & Higgins. They put rats at the junction between two compartments of a cage; one compartment was familiar, the other not; some rats were given a shock at the junction. The shocked rats chose the familiar compartment, the unshocked explored the strange area (as one would expect). Rats of both groups displayed what the authors interpret as signs of autonomic activity, namely, face-washing, grooming, scratching and trembling, but these were observed more in the shocked rats [310]. In neither of these examples is there the performance of one stereotyped act in place of another, as in the classical instances of displacement behaviour; instead there are seemingly irrelevant activities which accompany an adaptive change in autonomic function.

Another category of 'displaced' behaviour is found in acts which, though evidently in part substitutive, are not irrelevant to the situation in which they are performed. Barnett gives an example from the social behaviour of wild rats. As described in § 4.3.2, in artificial colonies containing females as well as males there is a great deal of fighting among the males and a high mortality, while in all-male colonies deaths are rare and fighting mild or absent. Yet direct observation of mixed colonies shows that there is no fighting for the females. In a typical sequence a female, whose subsequent behaviour showed that she was in pro-oestrus, was approached by one or more males, all of whom were rejected. On the withdrawal of the female to a nest box the males then resorted to fighting, although as a rule, in this colony, they were peaceful. Here is an example of behaviour which suggests a description in terms of 'overflow' or of 'release of tension'; but the behaviour was relevant to the situation to this extent, that each male was in his territory; and as we saw in § 4, a male wild rat does attack other males in his territory (though usually only strange males). Evidently the presence of females lowers the threshold for territorial fighting [19]. In this example, then, stimulation to perform one act (coitus) leads to the performance of another act (fighting): the combined effect of (i) the presence of a female soon to be in oestrus and (ii) the presence of another adult male is to increase the probability of fighting among the males. The two kinds of stimulation have an additive effect which is expressed in the intensity with which one kind of action pattern appears.

This kind of observation makes the notion of displacement behaviour which is completely irrelevant to the situation less acceptable. One way in which apparently substitutive behaviour can be brought about has been suggested by van Iersel & Bol, in a detailed study of stereotyped behaviour in terns (*Sternus*). The readiness to perform a fixed act, such as brooding, tends to inhibit others, such as preening or nest-building. The 'drive' for each act can be given a numerical value in terms either of a threshold value or of number of movements in a given period. When the values for two mutually inhibitory acts have a certain ratio, neither is performed, but a third act is. This is supposed to be a consequence of a reduction in the inhibitory effects of the first two. If, for instance, the drives for brooding and flight are at equivalent levels, neither occurs; but preening, released from the inhibitory effects of brooding, does occur [160].

This hypothesis, though without a physiological basis, may eventually help to account for a good deal of 'displacement' behaviour. An example is the perfunctory grooming of rats and other small mammals in situations in which several kinds of activity, including exploration and flight, are simultaneously provoked.

The extent to which we can properly speak of 'substitutive' behaviour, and in precisely what sense, will be decided only when the underlying physiology is known. The beginning of a physiological analysis, for processes which are at least analogous to some kinds of displacement behaviour, has been attempted in several species of mammals. The analogous phenomena are the 'psychosomatic disorders' of man [13]. In these an efferent output seems to be substituted for aggressive behaviour or for flight [17]; but the new output is not to the skeletal muscles but to the smooth muscle, for instance of lungs or gut, or to glands: the discharge is in fact autonomic. The medical aspect of the subject has been reviewed by Wolff [332]. Among the familiar psychosomatic disorders is gastric ulcer. French and his colleagues have studied the genesis of ulcers in monkeys (*Macacca*). They induced gastric secretion of hydrochloric acid by stimulating the hypothalamus with implanted electrodes. When this was done with the electrode in the posterior hypothalamus the peak of secretion occurred only after about three hours and the response was mediated by the pituitary and adrenal cortex; adrenocorticotrophin (ACTH) and adrenal cortical hormones produce a similar effect. When hypothalamic stimulation was repeated at four-hour intervals for many weeks some of the monkeys developed pathological conditions of the stomach, including ulcers [103, 104].

This work shows how the system of the hypothalamus, pituitary and adrenals can contribute to ulcer formation. The electrode in the hypothalamus may be regarded as a means of simulating the discharges, presumably from the cerebral cortex, which set the whole process going in human beings. This is supported by the work of Conger and his colleagues, who have studied rats which had been exposed to a 'conflict situation' in which they were shocked when they approached food: prolonged stress of this kind, especially if the rats were kept alone, led to gastric pathology and sometimes to ulcers [71].

It is uncertain just how these observations should be interpreted. Perhaps the ulceration was an incidental result of the autonomic and related stimulation which accompanies prolonged exposure to a stressful situation. In man at least, however, the origins of at least some psychosomatic disorders seem to be much more complex. Just as with 'experimental neuroses', discussed below, the extent to which comparisons may be made with our own species is not yet established.

In fact, the area of ethology formerly covered by the term 'displacement behaviour' is full of obscurities. There is a tendency now to see what was formerly called displacement behaviour as falling into two categories, neither of which corresponds to the original concept. (i) Some are held to be the result of the release of inhibition on some behaviour pattern which is consequently permitted expression even though the stimulation to perform it is not high. (ii) Others are by-products of stimulation, for instance an incidental consequence of autonomic activation. It has still to be decided whether a third category is required, of activities evoked as a result of some central facilitation. Certainly, in learning some such effect seems to play an important part (§§ 8.3; 9.5).

8.2.3 Experimental 'psychopathology'

8.2.3.1 'Experimental neurosis'. One of the circumstances in which 'displacement behaviour' occurs is when two incompatible behaviour patterns are simultaneously provoked. An example, already cited, is that of the male herring gull (Larus argentatus) in its territory, hesitating between flight from or attack on an intruder: the bird resorts to pulling at grass, an activity ordinarily performed when a nest is being built. This sort of behaviour is often said to reflect a 'conflict of drives' (in the example, the 'drives' are those for flight and attack). Such a situation does not, of course, always lead to any apparently substitutive behaviour: there may be vacillation or ambivalent behaviour. A male wild rat in his territory, faced with another, strange male, may oscillate between attack and amicable behaviour [19].

The experimental study of animals in conflict situations began with Pavlov. The whole field has been reviewed by Broadhurst [55]. Pavlov was concerned not with stereotyped behaviour, but as usual with learned habits. A CR may be set up to a stimulus such as an oval patch of light. The same animal may also be trained *not* to respond to another, rather similar stimulus, such as a circular patch of light. In later trials the difference between the two conditional stimuli is reduced, until the animal cannot reliably distinguish between them. Eventually, during this process, the behaviour of the animal changes: first, as would be expected, mistakes begin to be made; but, later, severe disturbance results: the ability even to make easy discriminations is lost, the animal resists being taken to the laboratory and it displays other atypical behaviour such as loss of appetite or excessive activity.

The name assigned to the abnormal behaviour is 'experimental neurosis', but it is uncertain to what extent it parallels any form of disturbed behaviour in man. Hebb [132] has examined this question critically. Most of the human conditions labelled 'neurotic' are displayed in a much wider range of conditions than is experimental neurosis: the latter is a response to a specific situation in which the animal has repeatedly undergone a traumatic or confusing experience. We need to know much more about the experimentally induced condition in laboratory animals and also human psychopathology (especially perhaps its physiology). It is surprising how little rigorous research is being carried out on these problems.

Behavioural upsets of the sort studied by Pavlov were ascribed by him to conflict between excitation and inhibition. This conflict still cannot be described in terms of neural function, but it evidently has its effect through the processes of arousal described in § 7.2.

The conditions in which experimental neurosis develops have been more fully displayed in experiments on trial-and-error (or instrumental conditioning). Cook strapped laboratory rats in a harness and trained them to flex a leg in response to a bright light; for this they were rewarded with food. If they flexed their leg in the absence of the stimulus they were punished with a slight shock. Once trained, they were further induced to ignore a dim light: if they flexed in response to this they were shocked but, if they did not, the light went out after five seconds. Finally, the intensity of the dim light was gradually raised, so that the discrimination was made more difficult. This procedure duly led to disturbed behaviour by the rats; but the greatest signs of distress were during the two stages before the discrimination was made more difficult, that is, when the leg flexions performed at the wrong time were punished [72]. In this kind of experiment we find, then, behavioural upset primarily due to a noxious stimulus in conditions from which the animal cannot escape. A major resemblance to the Pavlovian situation lies, in fact, in the confinement in harness.

Wolpe has suggested that disturbed behaviour is usually due to one of two kinds of stimuli applied *in confined conditions*. The first are noxious or pain-causing stimuli such as shock: these may be defined objectively as stimuli which usually lead to avoidance. The second kind are the ambivalent stimuli first studied by Pavlov. Ambivalence can be produced in three ways. (i) It may be due to making a discrimination difficult. (ii) It may result from increasing the delay before the reward is given: we have already seen that, during the interval between the application of the conditional stimulus and the reward, a major effort of inhibition may be required (§ 6.2.1). (iii) Finally, in some experiments, a similar effect is produced by rapid alternation of stimuli which tend to evoke opposite responses [333].

How confinement exerts its ill effects is not certain. It is easy to feel sympathetic to an animal in these experiments, and to reflect that a man too would become exceedingly disturbed in similar conditions. Indeed, experiments on human volunteers have shown that this is the case. Wolpe suggests that restriction of movement could act in three ways: (i) preventing escape may result in a cumulative action of stimuli which cause anxiety; (ii) anxiety may become associated with the surroundings in which the experiments are carried out; (iii) autonomic responses may become stronger in the absence of the outlet provided by activity of the skeletal muscles. The last suggestion is supported by the work of Anderson & Liddell [8, 199] on sheep (*Ovis*).

However, confinement is not essential for the production of experimental neurosis. This has emerged from studies of conflict situations in which the animal is free to move about. In most such situations, no disturbed behaviour results. Miller has given a general review [216] and, as he shows, much of the behaviour is of a kind easily predictable by common sense. The mythological ass starved to death when equidistant between two equally tempting bundles of hay; but in fact an animal faced with alternative goals, both attractive, quickly chooses one. By contrast, if it is between two dangers, and there is no third path for escape, it vacillates at the point equidistant between them. (It is not only cows that vacillate – or goats that are capricious.)

The most interesting effect is when a single place or action is made both attractive and a source of pain. Masserman [213] has done this with cats (*Felis*) and has produced typical experimental neurosis. He writes:

Masserman, the cat man,
Makes felines neurotic.
Are cats and humans
Truly asymptotic?

His procedure was to signal the presentation of food, for instance with a light, but to present with the food a strong puff of air which would ordinarily be a severe deterrent to cats. Among the effects of this treatment were not only disturbed behaviour but also internal changes which, as we see in the next sub-section, accompany anxiety state in man.

All these kinds of behaviour seem at first sight to be maladaptive. That is why they are labelled 'neurotic', 'pathological', and so on. But this is only an assumption, as we can see from the work of Maier on the development of behaviour fixations. Laboratory rats were faced with an insoluble problem in a Lashley jumping apparatus (figure 36, page 108). Whatever solution they chose they were punished at random, that is, on half their jumps on the average. They were forced to jump by means of a blast of air (now realized to be a noxious stimulus). The rats tended to develop a fixed response, such as jumping always to the right: indeed, they persisted with their fixated behaviour even when the situation was changed, for instance if the door were opened and food visible on the left-hand side. An important feature of the behaviour was that, once a fixation had developed, the rats were less resistant to making the jump [207]. The detailed interpretation of this study is uncertain, as Russell & Pretty [266] have shown. However, a behaviour fixation may be regarded as an adaptive response to a situation in which the main reward has come to be, not the attainment of the food behind one of the doors, but avoiding the noxious stimulus. The fixation also makes possible the release of excitation through muscular action – a release which, as we have seen, is not available to animals in the situations in which experimental neurosis develops.

8.2.3.2 *Physiology*. Expressions like 'release of excitation' are apt to arouse derision among physiologists whose work on the nervous system is based on the very precise measurement of clearly defined electrical and chemical changes. Of course, the behaviour described in the preceding pages can be analysed rigorously and quantitatively, up to a point, but its physiological basis is still obscure. There is irony in this. Pavlov was a physiologist; his aim in studying behaviour was explicitly to separate it from metaphysics and from the use of terms like 'will', 'emotion' and 'consciousness'; he regarded his studies of CRs as essentially an inquiry into the physiology of the mammalian forebrain. But his work was done long before the time when any satisfactory account of neural function in the CR or in experimental neurosis could be attempted.

Today, more than half a century after Pavlov began his work on CRs, it is possible to say a little about the internal changes which accompany experimental neurosis and allied states. In doing this we are following in the path of Darwin as much as Pavlov: we are studying – as Darwin urged – the physiology of the 'emotions'.

In human terms the main relevant emotions are fear and anxiety. These, reported by volunteers during laboratory experiments and by patients in consulting rooms, are accompanied by measurable signs: among the signs are sweating, tachycardia, raised blood pressure, blood glucose and pulse and respiratory rates, erection of the hair and dilated pupils; these reflect activation of the sympathetic nervous system, while evacuation of the bladder and rectum, especially associated with anxiety, which may also occur, indicate parasympathetic discharge. All these may be observed and measured in other mammalian species: Masserman [213] found them in his 'neurotic' cats. In a given individual (human or not) either sympathetic or parasympathetic activity may predominate at a particular moment but, as we shall see (§ 8.3), arousal is particularly accompanied by sympathetic activation.

The pioneer in this field was W. B. Cannon, whose account of the activity of the sympathetic nervous system and of the adrenal medulla, together with their effects in the body, has long been familiar. He emphasized that the bodily changes in 'pain, hunger, fear and rage' are adaptive: they increase the animal's capacity for violent action, such as is needed in flight from an enemy or in attacking prey. This obviously holds for the increased blood supply to the skeletal muscles, the higher breathing rate and the raised blood glucose. Further study has added complexities to Cannon's scheme [111, 112]. The sympathetic and parasympathetic systems may be distinguished anatomically and also by the different sets of drugs which imitate their actions; but they do not always seem to act in opposition to each other. The fact that in an 'emergency' an animal may defaecate and urinate in an atypical way shows, as we saw above, that parasympathetic excitation may occur in situations where sympathetic activity is increased. (Perhaps the discarding of faeces and urine before violent action is an advantage.) More certainly, other consequences of parasympathetic action are useful: vagal stimulation results in release of glycogen from the liver, and this increases the ability of the animal to maintain a high level of blood glucose.

Changes of this kind have been observed, not only in experimental neurosis, but also in the state of acute distress induced in wild rats by attack, in a strange place, by another rat (§ 4.3.2). In the conditions in which these rats were observed, death is the rule after repeated attack, sometimes only after some days but often within a few hours. The strangest obvious feature of death from this cause is that it usually occurs in the entire absence of wounding; severe wounding was never observed. Barnett and his colleagues have shown that the attacked rats tend to be hyperglycaemic but that their liver glycogen is low [29]. This fits the picture of the mammal in an emergency already familiar from other studies.

Death under attack, although usual, is not invariable. A few rats adapt themselves to a subordinate position and survive in good health: this can happen even when the dominant rat is exceptionally fierce [19]. Why there should be this difference is not known; but there is evidence from several sources of differences in 'personality', related to different kinds of autonomic response during stress.

Pavlov himself described different kinds of temperament in his dogs, reflected in the form taken by their 'neuroses'. In general, in experimental neurosis the behaviour may suggest (i) a state of continuous fear or anxiety, marked by unresponsiveness, trembling and crouching, and decreased gastric function; (ii) chronic rage, marked by restlessness, agitation, aggressiveness and increased gastric function. Funkenstein [108] and Gellhorn [112] among others have suggested that these 'emotional' states can be distinguished physiologically. In a given individual there is a parallel between the physiological changes in fear and the effects of administering adrenalin. These effects have been familiar, from the work of Cannon, for several decades. However, since 1948 it has been known that the adrenal medulla also secretes a related substance, noradrenalin, which has the special effect of causing vaso-constriction by a local action on small blood vessels. The physiological changes in anger parallel those of injecting noradrenalin. Evidently, in a given individual, one of the two hormones is dominant. Funkenstein also refers to differences between species in this respect: he claims that aggressive animals such as lions secrete much noradrenalin while very social and restrained animals, such as baboons, have a higher adrenalin output. There is evidently a big field here, for combined studies of behaviour and physiology.

However, the autonomic activation and secretion of adrenalin and noradrenalin constitute only a fragment of the complete syndrome of changes which take place when bodily homeostasis is put under severe strain. A variety of conditions ('stressors'), including starvation, exposure to cold, infectious illness, poisoning, burns and forced exercise, are accompanied by increased secretion of adrenocorticotrophic hormone (ACTH) by the pituitary and consequent raised output of steroid hormones by the adrenal cortex. If the hostile conditions persist the adrenal cortex enlarges through hypertrophy of its cells. This is one means by which the mammalian body is maintained in different circumstances [111]. These changes occur also in anxiety and allied states. In wild rodents, as well as in laboratory rats and mice, crowding and the consequent intraspecific conflict are accompanied by enlargement of the adrenal glands [68]. Barnett found that wild rats which had been for long in conditions of conflict had enlarged adrenals; moreover, rats which had just been under severe attack usually had adrenals drastically depleted of their reserve of hormone precursor (plate 14). Barnett therefore used the term 'social stress' to denote the disturbance induced in this way [22]. Thus the autonomic activity and adrenalin secretion studied by Cannon in 'emotional' states are followed by a further syndrome, of which the state of the adrenal cortex provides a convenient index. Further, this whole series of neural and endocrine responses is part of the lethal disturbance which occurs in social stress.

Whether the same dual process occurs also in experimental neurosis seems not yet to have been studied. Mirsky and his colleagues have, however, subjected monkeys (*Macacca*) and rats to conditions arousing fear; ACTH was found to reduce the effect of the frightening stimulus in both species. They suggest that secretions of the adrenal cortex either reduce anxiety or make the bodily response to anxiety more efficient [221]. Similarly, Levine & Soliday have brought evidence that ACTH reduces the rate at which rats learn to avoid a noxious stimulus [197]. There is scope for a great deal more work on these lines.

The autonomic and endocrine responses to an emergency are still only a part of the total bodily activity. In accounts of the physiology of 'emotion' they have been given much prominence, partly no doubt because exact, quantitative information exists for them; another reason is that animals in laboratories are often restrained, if not in harness at least in a small volume of space, and so are unable to react with the full resources of their somatic muscles. Yet the most obvious result of the approach by a predator or a rival, or of pain or other noxious stimulus, is usually some vigorous action of the whole animal, namely, flight or attack. This is accompanied by autonomic activation and its associated endocrine secretions; these help to maintain the activity, and to restore the body when the animal can rest again; but the peak of sympathetic activation is soon passed unless the animal is under restraint: in the latter case, experimental neurosis, behaviour fixations or ulcers and so forth are likely to develop, evidently as a result of a non-adaptive persistence of internal states which are useful only in emergencies.

The whole of the assembly of responses is, of course, set going by processes in the brain. A little is known about the rôles played by different parts of the brain. If a mammal such as a cat or dog has its cerebral cortex surgically removed it responds to quite mild disturbance with a violent expression of 'rage' or 'fear'. Thus this pattern can be evoked provided that the diencephalon (which includes the thalamus and hypothalamus) is intact. Evidently one aspect of cortical function inhibits this kind of response. This, however, tells us little of how attack or flight (with their accompanying autonomic activity) are set going in the intact animal. Still less can we account physiologically for experimental neurosis or death under social stress.

The intensity of this sort of excitation can be measured in terms of the activity of certain regions of the cerebral cortex. These regions are the sources of fibres passing to the reticular formation in the brain stem (§ 9.5). The greater the stimulation of these areas, the greater the behavioural arousal of the animal. There is a sequence of behaviour from curiosity or attention, through fear or anxiety, to terror, which parallels the neural activation [237]. (Evidently, death from 'social stress' occurs in rats at the extreme of the scale of terror.) It follows that a moderate degree of activation favours effective behaviour, but beyond a certain point behaviour begins to become disorganized. Here again is an example of an optimum in the level of stimulation (§ 7.3.2.1). This concept is evidently of the greatest importance, not only in the study of mammalian behaviour and physiology but also for the understanding of human behaviour. Much more work ought to be done on it.

8.3 'GENERAL DRIVE'

Some emphasis has been given above on the plural nature of the processes underlying 'drive phenomena'. Nevertheless, there is much evidence that the internal processes which impel one kind of behaviour can often also make the performance of another pattern more likely. In § 8.2.2 there was the example of rats which resort to fighting when sexually stimulated but prevented from copulating. This sort of observation leads inevitably to the notion of *general drive*: that is, internal processes which tend to impel an animal to action, though not to any specific action. Such a notion has been reached in psychological laboratories, independently of the work of zoologists studying displacement behaviour in the field or in quasi-natural conditions. Hull [158], in particular, referred to 'generalized drive' in describing certain effects of states such as hunger and thirst on the performance of learned habits. An experimental example is provided by the work of Webb. A habit is learned under the influence of hunger; will the learned act be performed under the influence of another need, such as thirst? And, if so, will the strength of the response vary proportionally to the strength of the new need? Webb used as his criterion the number of trials required to produce extinction of a habit learned under hunger. Rats deprived of water (but not food) for three hours before testing took nearly twice as many trials to extinction as rats which were neither hungry nor thirsty; and a further group deprived of water for twenty-two hours took about two and a half as many trials as the controls. Other work has confirmed Webb's observations; Spence [298, pages 189 e.s.] concludes that Hull's 'hypothesis of a general drive factor to which any and all acting needs contribute' is supported.

Allied to the notion of general drive is that of the additive effect of drives. We may ask the general question: is response strength a function of the combined needs of an animal at a given moment? On this the experimental facts do not present any simple picture. For instance, Kendler trained hungry rats to press a lever for food; and he found that trained rats which had been deprived of both food and water for twenty-two hours pressed the lever *less* persistently than similar rats which had been deprived only of food for that period. This did not, however, hold for shorter periods of water deprivation combined with the same degree of hunger [168]. It is a familiar fact of human experience that thirst interferes with eating; Kendler's work shows that the same applies to rats and suggests that a simple arithmetical approach to 'drives' and response strength is often inadequate.

Observations on additive effects of noxious stimulation are still more interesting [298, pages 186 e.s.]. Several studies have shown that giving rats a mild shock before presenting food or water increases the amount consumed. This suggests an additive effect of 'drives', since the shock is liable to induce flight. However, the effects of noxious stimulation on learning and learned habits are not so simple. We saw in § 7.3.2.1 that learning does not, beyond a certain point, improve in proportion to increasing drive strength: it is in fact sometimes possible to demonstrate the existence of an optimum drive level for a task of given difficulty; (compare § 8.2.3.2 for optimum level of arousal). To quote Spence: 'If conditioned emotional (fear arousing) stimulus cues are present in the test situation, overt responses (crouching, face washing, escape responses) that compete with the criterion (drinking, eating) response occur' [298, page 195].

There is evidence, then, that processes which impel different kinds of behaviour can sometimes act in an additive way in determining the net amount of an animal's activity. This notion has been recently given new life as a result of studies of a previously neglected part of the brain. The brain stem and thalamus in a mammal contain not only centres and nuclei related to specific functions and individual cranial nerves and tracts, but also a matrix consisting of large numbers of small cells. This,

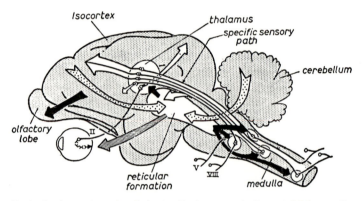

61. Reticular formation of cat's brain. Dark arrows indicate inhibitory effects of the reticular substance; the hatched arrow represents facilitatory or inhibitory action; dotted arrows show tracts to the reticular formation. The diffuse projection system is not shown. (After Hernandez-Peon [141].)

now called the reticular activating system (RAS), is further described in § 9.5. The RAS receives collateral fibres from the major sensory pathways. The latter pass, by way of synaptic connexions in the thalamus, directly to the cerebral cortex, and constitute the *specific* projection system. Impulses in the collaterals, by contrast, cross many synapses in the RAS, but they too eventually reach the cortex; these pathways form the *diffuse* projection system, so named because the reticulo-cortical tracts end in many areas of the isocortex (figure 61).

Injury to the RAS causes coma or somnolence. Stimulation causes awakening and a state of alertness which can be studied either by direct observation of behaviour or indirectly, for example by means of the EEG. 'Emotional' states are accompanied by a high level of arousal in the sense of activation of the RAS; and there is a correlation between activation of the RAS and that of the sympathetic nervous system. In the RAS, then, we have a part of the central nervous system on which all kinds of activity depend. Any of a great number of external stimuli can lead to arousal; and a variety of internal conditions, such as low blood sugar, can do the same. A given *specific* stimulation activates in the first place a pattern of motor output which is (whether as a result of a learning process or not) appropriate to the situation. But sometimes the appropriate pattern cannot be performed. In that case it is possible that the general state of arousal, perhaps through widespread facilitation or summation in the cortex, leads to the performance of another act instead. This possibility requires further study.

8.4 ASPECTS OF ONTOGENY

8.4.1 *Sensitive periods*

The readiness with which certain kinds of behaviour are performed or learnt varies in two major ways: first, it is influenced by brief fluctuations in the animal's state; second, it alters with age. We now turn to the effects of age. On this some of the most precise and detailed observations have been on nidifugous birds: recently-hatched moorhens (*Gallinula*), coots (*Fulica*), ducks (Anatidae) and others readily learn to follow any moving object of suitable size, but they soon lose this propensity. There is only one, sometimes brief, stage or sensitive period in the life history when this *imprinting* can be easily established [143, 153].

These are instances of the learning or development of a stereotyped activity, but learning ability in general also changes with age. In laboratory rats the ability to learn mazes, discriminations and puzzle boxes reaches a peak at the age of about nine weeks, after which it declines [233, page 351]. The details of these changes are complex, Stone, for example, on the basis of a substantial series of experiments, concluded that from the age of 70 days learning ability remains constant for many months. But he did find important differences with age: for instance, previously learned habits were more easily discarded by young rats; also, they took fewer trials than old ones to learn the most difficult tasks [303, 304].

However, the obvious example of a sensitive period influencing learning in rats was given in § 7.4. We saw there that diversity of experience in early life is necessary for the development of maximum learning ability. From this point of view a laboratory rat reared in a small cage constitutes an 'experiment', by contrast with one brought up in a complex environment. A wild rat or other animal in its ordinary environment is inevitably exposed to a great diversity of conditions: what is abnormal is the impoverished environment inflicted on most laboratory rats.

8.4.2 *The need for stimulation*

The preceding section refers to two general features of the ontogeny of behaviour: first, there are sensitive periods for the development of of abilities and of stereotyped behaviour patterns; second, optimum development requires plenty of external stimulation in early life. The same principles are illustrated by recent work on the effects of stimulation of the skin in rats and other mammals.

The importance of cutaneous stimulation in social behaviour is discussed in § 4.2. Much work has also been done experimentally on the effects of stroking the skin of young laboratory rats [25, 196]. The animals, stimulated in this way or merely by handling for a few minutes each day, grow more quickly than controls which are left alone. There is no difference in the amount of food eaten but the experimental animals evidently make better use of the food they do eat. There seems to be a generally higher rate of development: for instance, the cholesterol content of the brain, and hair growth, are both increased. These effects on growth are accompanied by an improvement in learning ability. Handling has this effect especially if it is done before the rats are twenty days old: evidently there is a sensitive period when they are most responsive to this sort of stimulation.

There is an obvious parallel here to the effects of environmental diversity on learning ability. It is, however, doubtful whether this parallel signifies that there is a similar underlying physiology. It seems more likely that the physiology of the effects of handling will emerge from studies of resistance to stressors in handled animals. Handling, especially during the first few days of life, reduces the ill effects of wounding, of prolonged deprivation of food and water and of exposure to cold, among other things. A correlate of these observations is that handling reduces 'emotionality': that is, a handled animal, on being put in an unfamiliar open space, explores confidently and shows little sign of exceptional autonomic activity; a non-handled animal tends to freeze, and to defaecate and urinate excessively.

It is possible that handling itself constitutes a mild stressor: that is, it activates the hypothalamic-pituitary-adrenal system which, in the adult, plays a major part in the adaptive response to cold, injury and so on. In the absence of stress, handled animals have the same output of adrenal cortical steroids as non-handled controls. If they are then subjected to pain or poisoning, steroid output rises in both groups, but is higher in the animals which have not been handled; this perhaps reflects the greater susceptibility of the controls to the ill effects of the stressor. However, steroid output during the first fifteen minutes after a severely painful stimulus is higher in *handled* animals. Evidently, it is delay in making the endocrine response that is maladaptive. It is as if the ability to respond effectively to stress in later life is brought to a higher level by mild early stimulation.

Another view of this work is suggested by the observation of Lát and his colleagues. The growth rate of a rat throughout its life is greatly influenced by the amount of food it receives during its first two post-natal weeks: if it is reared in a small litter (say, of three) its weight at weaning is greater than it would have been if reared in a litter of twelve; and this difference persists at all later ages. Differences in growth are paralleled by differences in behaviour: fast-growing rats are more active and more 'intelligent' (§ 6.4) than slow-growing [192]. Much remains to be done on the analysis of the behavioural differences. But, since an effect on growth also influences behaviour in a manner resembling the action of handling, it seems that metabolic processes common to both experimental procedures should be sought.

8.5 CONCLUSIONS

Observations on drive or motivation have been described in every chapter of this book. An attempt will now be made to give a summary of the whole subject.

1. In a constant environment there are changes in the ease with which stereotyped forms of behaviour can be evoked. Similarly, fluctuations of internal state influence both the rate at which tasks can be learnt and the readiness with which learned habits are performed. Further, great persistence may be shown in achieving a consummatory state, such as a full stomach, with its consequent ending of internal stimuli which provoked the initial activity. All these features of behaviour may be termed 'drive phenomena'; they come under the conventional heading of 'motivation'.

2. The internal states and external agencies which make an animal active are usually related, directly or indirectly, (i) to the maintenance of steady states in the individual (that is, they influence homeostasis), or (ii) to reproduction. This is to be expected. The behaviour of members of any species, like their other features, is a product of natural selection; accordingly, as a rule it has survival value.

3. Nevertheless, it does not follow that all behaviour is directly and immediately related to survival or reproduction. (i) Some behaviour is vigorously performed although it is of no immediate use. This is the case with exploratory behaviour; the value of such behaviour in the

adult depends on subsequent need for topographical knowledge; in the young animal exploration is also important for the development of learning ability. (ii) Some behaviour which is only preliminary to the performance of a 'useful' act has 'reward value': that is, it may reduce the readiness with which a stereotyped sequence (such as courtship) is performed; or it may provide reinforcement in a learning experiment. An example is coitus without ejaculation (§ 7.3.2.3). (iii) An animal put experimentally in an entirely unnatural situation, such as one in which it can stimulate its brain by pressing a lever, cannot be expected to behave homeostatically, since its ancestors have not been subjected to natural selection in such situations and so have not evolved equipment to deal with them adaptively.

4. Much of the analysis of drive depends on physiology. The best example at present is feeding behaviour (§ 3.5). The internal factors which influence the nervous system include, among others, the composition of the blood, the degree to which the stomach wall is distended and impulses from the sense organs of the mouth and pharynx. In the brain the cortex and thalamus, the hypothalamus (with its 'centres' for both hunger and satiety) and the reticular formation must all play a part; but, except perhaps for the hypothalamus, little is known about how the brain works.

Just as there is a multiplicity of mechanisms determining the intensity, character and cessation of feeding, so there is sometimes a lack of correlation between different components of feeding behaviour: food-seeking activity, for instance, does not necessarily alter in parallel with actual eating (§ 3.4). These conclusions apply quite generally: they can be readily exemplified also from reproductive behaviour.

5. When a situation requires a vigorous response (attack, flight, coitus) the somatic output is accompanied by changes in the autonomic nervous system and endocrine organs (especially the pituitary and the adrenal medulla and cortex); these help to maintain the body when heavy demands are put on it. The whole constellation of visible changes constitutes the 'expression of the emotions'. Most of the visible signs have a straightforward physiological significance; but some may also act as signals to other individuals.

6. The full development of the ability to withstand hostile conditions (stressors) probably depends on exposure to mild noxious stimulation in early life. The development of maximum learning ability certainly requires early exposure to a complex environment. These are examples of the importance of sensitive periods in ontogeny: during such periods the ability to develop or to learn certain kinds of behaviour is at a

maximum; if the appropriate external conditions are absent at the critical time the result may be a permanent behavioural deficiency.

7. Sometimes, when a vigorous response is aroused, behaviour seems to be ill-directed or positively disadvantageous. This applies both to 'displacement behaviour' (which may be observed in natural conditions) and to 'experimental neurosis'. Both these terms have been applied to rather heterogeneous categories of behaviour. (i) Sometimes, seemingly abnormal behaviour is in fact adaptive, as when a behaviour fixation enables an animal to avoid pain; or when a displacement act functions as a social signal. (In the latter case the act is said to be 'ritualized'.) (ii) An activity, apparently futile, may be a by-product of a useful physiological process, such as autonomic activation. (iii) There remains an enigmatic group of phenomena in which it seems that stimuli which usually evoke one activity lead instead to the performance of another; the substituted activity can be of varying degrees of relevance to the animal's circumstances. Perhaps this group will eventually be explained in terms of release of inhibition or of facilitation in the central nervous system.

8. The capacity for performing activities (including 'play') beyond those demanded by the immediate situation may confer a long-term advantage on a species. A reserve of 'energy' for exploration or for substitutive acts could be a source of behavioural preadaptation to new environments in the same way as heterozygosis provides a pool of genetical adaptability.

9. The fact that drive phenomena have a plural physiological basis does not preclude the possibility that there is some form of 'general drive': that is, that there are central nervous processes which influence an animal's net level of activity, regardless of the particular form that the activity takes. The facts of interaction between one pattern of behaviour and another, found both in stereotyped behaviour and in learned habits, have suggested this hypothesis. The concept of general drive has also been related to the rôle of the reticular activating system of the brain stem in maintaining arousal through general facilitation in the cerebral cortex. In this, as in most other aspects of the subject, nearly everything remains to be discovered.

9

Brain and Behaviour

When all is said and done, the fact remains that, for the beginner, the understanding of the brain's structure is not an easy thing. It must be gone over and
forgotten and learned again many times before it is definitively assimilated by the
mind. But patience and repetition, here as elsewhere, will bear their perfect fruit.

WILLIAM JAMES

9.1 THE RÔLE OF THE BRAIN

Of all physiological facts, those concerning the brain seem most likely
to help to explain behaviour; and, although this is a truism, we may now
ask why. The central nervous system is the organ through which the
sense organs act on the effectors. Hence, although all other organs may
influence behaviour, the CNS and the nerves carry especially large
amounts of information to the muscles and glands. While, for example,
the blood enables the muscles to remain operational, by carrying substances to and from them, the nervous system determines the moment,
duration and intensity of the contractions of each muscle and also their
relations to the contractions of other muscles. One of the most remarkable features of neural action is that it can impose a *pattern* of activity on
dozens of muscles and many millions of muscle fibres.

This patterning exists even in a simple reflex, as C. S. Sherrington
(1859–1952) described more than half a century ago. He wrote: 'a great
principle in the plan of the nervous system is that an effector shall be at
the behest of many receptors, and that one receptor shall be able to
employ many effectors' [289]. This arrangement, fully described in
any modern neurophysiological text, makes possible the interaction
of many afferent inputs to produce the animal's pattern of activity. But
it is not only the incoming signals that interact: there are also patterned
activities within the CNS that influence behaviour. These systems may
be called 'representations' of the external world of the animal [338].
They may be characteristic of the species, in which case the behaviour
they determine is of the kind we have called stereotyped: the representations will then be of releasing stimuli. Or they may be the result of the
adaptation of the individual to its own particular environment, that is, a
consequence of learning.

One can talk in only a vague way of 'representations', because we do not know the physical basis either of fixed action patterns or of the more plastic kinds of behaviour. But one can be more precise about the input to which the brain is continually responding in the awakened animal. Information is unceasingly fed into the brain from the external sense organs which give a picture of the outside world; and also from the internal sense organs which give a picture of the state of the body from moment to moment. The proprioceptors make an especially large contribution: almost half the axons of the nerve to a skeletal muscle are sensory, and most of these are proprioceptive; that is, they register deformations and stresses in the moving parts (muscles, tendons and joints) [200]. Injury to the tracts of proprioceptive fibres destroys the fine pattern of movements.

Two general principles may be stated concerning the input. The first is that, as we infer from behaviour, only certain patterns of stimulus evoke a specific response in the CNS: the remainder of the input is somehow prevented from having any special effect (although, as we saw in § 8.3, it may have a *general* effect of arousal). Out of the great mass of information reaching the brain, only certain patterns have a 'meaning' for it, because only those patterns correspond to a representation in the brain. Lashley gives a good, if complex, example. He describes how his early efforts to train laboratory rats to respond to visual stimuli led him to conclude that the animals were blind. 'Then by chance we hit upon the method of having the animals jump against the objects to be distinguished. This . . . put the question in a way that was intelligible even to rats' [188].

We saw in § 5 (though in different terms) that a representation may depend to a large or a small extent on individual experience. Patterns of input which do not correspond to a representation are ignored. This may be because the animal never in its life possesses the central nervous equipment to deal with them: in stereotyped behaviour it is commonly found that the stimulus patterns which evoke response are highly specific (§ 5.1.1). But often, environmental features have no effect only because habituation has taken place (§ 6.2.2). The brain is an organ which selects the inputs to which the animal will respond as well as the form which the responses take.

The second principle is that a great deal of central nervous function involves negative feedback. This term applies to any system in which a change tends to be nullified by its bringing about another change proportional to itself. Thermostats, governors of steam flow and the servomechanisms used to determine the direction of guided missiles are examples from engineering. Some of the consequences of this sort of arrangement in the body have been given in § 3.4 in the discussion of the control of feeding. Insofar as behaviour is homeostatic, it is based on negative feedback. All movement, and the maintenance of posture and spatial equilibrium, demand it: departures from an animal's orientation to gravity or to particular objects must be quickly corrected; in this the proprioceptors play their essential part. The scope of this principle has been impressively illustrated by the work of von Holst & Mittelstaedt [155]. Their work has been largely on insects, but the principles they propose apply generally. A change in an animal's environment evokes a corresponding adjustment in the animal: for instance, it may move its eyes to keep them aimed at a moving object. But, if the eyes are themselves moved by the action of the animal, so that the environment ought to *appear* to move, no compensating movement is made. Subjectively, if a man turns his head he does *not* get the impression that his surroundings are whirling around him; but, if he closes one eye and pushes the other sideways with his finger, the things he can see do appear to change position.

In performing all these functions the CNS is very unlike most machines; it is therefore instructive to examine the problems of contriving a machine which would carry out the functions of a simple nervous system. Consider the recognition of shape. If one wished to devise a machine which responded in different ways to a number of different shapes one would first think, perhaps, of an array of photo-electric cells, as a parallel to the light-sensitive cells of a retina, linked to a computer, or 'brain'; but as soon as we consider what a retina and brain actually achieve, such crude comparisons fail. First, in an animal such as a rat [133] or an octopus [338], recognition does not depend on the position of the image in the visual field: provided the receptors are activated in a specific pattern it does not matter which receptors are in use. Secondly, the size of the object relative to the visual field, and the angle at which it is presented, can vary over a wide range; this is the phenomenon of stimulus generalization described in § 7.3.1.3. To devise a machine with only these properties would be a formidable task, though something of the sort has been attempted [203]. But, finally, in a rat, as we shall see below, removal of a substantial part of the region of the brain (the striate cortex) especially concerned with vision does not necessarily destroy these abilities.

One consequence of this complexity of brain function, and of its lack of similarity to any of our familiar machines, is that we have no satisfactory language in which to describe how it works: it is difficult even to formulate in detail the questions which need to be answered. This difficulty is made more clear if we examine the methods available for studying the brain.

9.2 METHODS OF STUDY

9.2.1 *Anatomy and gross injury*

If one wishes to understand the brain, a first step is to study its structure, both gross and microscopic, in a given species or, better, in a series of species. The brain of a rat is illustrated in figure 62. It is typical of a small mammal.

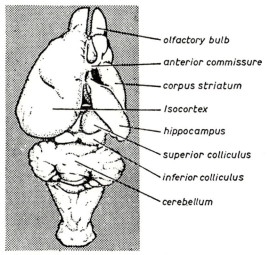

olfactory bulb

anterior commissure

corpus striatum

Isocortex

hippocampus

superior colliculus

inferior colliculus

cerebellum

62. Brain of rat: dorsal view with dorsal cortex and corpus callosum removed on right. (After Beach [36].)

The brain stem (medulla, pons and midbrain) is, like the spinal cord, largely an assembly of 'reflex centres': it receives the inputs of the cranial nerves, except the olfactory and most of the optic, and contains the origins of the motor fibres in them. Particular regions are associated with specific reflex acts such as lachrymation, with more complex patterns such as breathing and (in the midbrain) with postural adjustments. There is also a large cerebellum of which the function is evidently the control of the details of movements: it is the cerebellum which (as far as we know) makes the proprioceptive feedback effective. In all this the mammalian brain resembles that of other classes of vertebrates.

One feature of the brain stem is of especial importance in understanding the working of the forebrain: this is the reticular activating system. Its description will be deferred till later (§ 9.5).

The most distinctive feature of the mammalian CNS is the increase in relative size and the elaboration of the forebrain. (The only part of the forebrain which largely retains its primitive rôle is the hypothalamus, on which a good deal has already been said in § 3.4.) It is usual to emphasize especially the vastly enlarged cerebral hemispheres (figure 63):

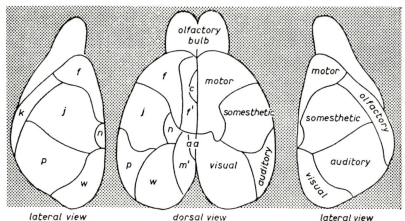

63. Brain of rat: the maps show histologically different areas (marked with letters) and the supposed 'functions' of some of these areas. (After Lashley [187].)

these have an outer layer of cellular tissue, the cerebral cortex, while most other vertebrates retain the cell bodies within the tracts of fibres (white matter) which connect one part of the CNS with another. The working of the cortex, however, cannot properly be discussed without referring also to the thalamus which too, in a mammal, is especially large and complex: it is in the thalamus that many of the afferents to the brain make their first synaptic connexions; from there impulses are relayed to the cortex (figure 74, page 234). Thus the cortex has converging upon it the main input from the special senses. Also large and complex – though until recently they have received less attention – are structures at the base of the forebrain which have sometimes been called the basal ganglia or (under the impression that they are concerned with smell) the rhinencephalon. They are now often called, with parts of the cortex, the limbic system (figure 72, page 228).

We have only the dimmest understanding of how all these structures work; but it is at least becoming clear that the cerebral cortex, the thalamus, the limbic system and the reticular activating system are all in continuous interaction and that no one can be understood without the others.

This assemblage of interacting structures is certainly responsible for the complexity of mammalian behaviour. It represents a concentration of overriding control at the front of the brain, reflected in the enlargement of the forebrain already mentioned. In its evolution, the forebrain originated as a purely olfactory receiving station; but, during the rise of the mammals, and also on a number of lines within them, the olfactory influence has become progressively less. This is illustrated by the fact that the output to the skeletal muscles originates in a particular part of the cerebral cortex (in man, the pre-central gyrus): from here the pyramidal tracts run, to make direct synaptic connexion with the motor neurons of the ventral horns of the spinal cord; from the latter pass the fibres which end in the motor end-plates of the muscles themselves. In mammals there is consequently a unified and very direct control of behaviour by the cortex and its associated structures. This feature of the mammalian brain is expressed in Sherrington's term 'the final common path'.

Within the mammals the complexity of learned behaviour can to some extent be related to brain size. This, of course, does not hold in any simple way, or large whales (Cetacea) would be five times as intelligent as man. Moreover, as Anthony [10] has shown, the significant relationship is not even weight of brain in relation to body weight: if it were, marmosets (*Hapale*) would be more intelligent than any other primates. In fact, the important parameters, in relation to intelligence, are of a different kind. In the most primitive of extant eutherian mammals, the Insectivora, the forebrain is still largely dominated by the olfactory input; most of the non-olfactory cortex receives afferents from the other exteroceptors. In the primates, at the other end of the scale, there are large cortical regions, poetically called the silent areas, which are not functionally related to any special sense; nor are they motor (figure 64). It is the progressive reduction of olfactory influence on the cortex and the increase in silent areas which are the main concomitants of behavioural complexity in a mammal.

One other general, quantitative feature of the brain is obviously important: the numbers of cells in the forebrain, especially the cortex, are vast, and their connexions are bewilderingly elaborate. Each neuron is in synaptic connexion with hundreds of others; and there are arrangements between the cells of different cortical layers which resemble the feedback loops of engineers. It is not easy to see what use all these cells are, especially since an adult mammal may still perform well (as we shall see later) when deprived of a sizeable proportion of them. Hebb [133] has plausibly suggested that the great mass of connecting neurons in mammals is especially concerned with early deutero-learning (§ 7.4); this, as we know, is of outstanding importance in the development of intelligence. However, the facts on the rôle of cortical tissue in infancy are conflicting.

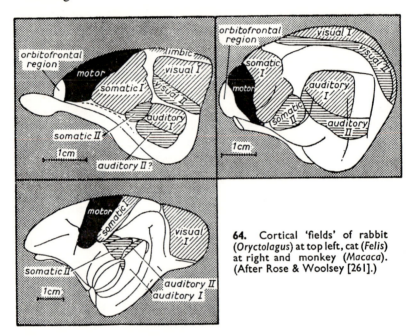

64. Cortical 'fields' of rabbit (*Oryctolagus*) at top left, cat (*Felis*) at right and monkey (*Macaca*). (After Rose & Woolsey [261].)

9.2.2 *Injury and 'localization'*

The study of structure, gross or microscopic, gives only the beginnings of knowledge of brain function. One can admittedly learn much of the connexions of sense organs and receptors to groups of nerve cells, and also of connexions within the CNS. Gregory has compared the pictures derived in this way to the blueprints used by engineers: they provide a lay-out of the components, but no information on their functions. It is (he adds) of course possible to say something of the impulses which pass in the tracts, and of the results of interrupting them. Such physiological accounts of the brain may be compared to a wiring diagram: they still do not say what the parts do, only how they are connected [in 313].

By the accident that clinical information was available before the results of planned experiment, early notions on localization of function in the brain have been derived largely from study of the effects of injury. This is like trying to find out how a very complicated and delicate machine works by taking a hammer to bits of it at random. The method, together with some more refined procedures, has, however, yielded the familiar maps of the cerebral cortex (figures 63, 64, 65). These maps show the main projection areas and the motor area. There also exist,

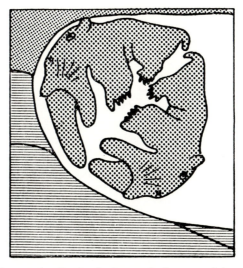

65. Sensory and motor localization in the cerebral cortex of the rat. On the left the pattern of representation of motor function in the pre-central area; on the right, the sensory map – an approximate mirror image of the motor. (After Woolsey [334].)

for some species, quite detailed maps of the somatic sensory region, in which the surface of the body is represented in a distorted form corresponding to the relative importance, as sources of information, of the various districts of the skin: the snout in a pig and the finger tips in a man have enormous representations. The main motor area, too, can be mapped; here the representation of the somatic musculature reflects the complexity of the movements performed: in man muscles of the larynx and fingers have especially large regions. All such maps are now regarded as showing the points of arrival and departure of the main, specific afferent and efferent tracts: they tell us little about what the cortical neurons, in all their variety, are doing.

Behaviour after brain injury has of course been studied in rigorous conditions in the laboratory, especially in rats and monkeys, where it has often led to valuable observations. Examples were given in § 3.4, where the effects of exceedingly small injuries to the hypothalamus were described. Others will be found below. Most of the experiments involve making roughly symmetrical bilateral lesions. Rather little is known about the effects of injuries confined to one side in laboratory animals, except that often it is difficult to detect any change due to the lesion.

9.2.3 *Electrical and chemical methods*

Probably, advances at the cellular level will be needed for major progress in understanding the brain. The great achievements of cellular neurophysiology so far have been in the analysis of the nerve impulse, that is, the electrical and chemical changes in axons and at synapses. This work has been done largely on peripheral nerves and autonomic ganglia. We know rather little of what happens in the brain, and hardly anything with certainty of the changes in nerve cells that constitute learning. Synaptic connexions become more efficient with use (at least in the spinal cord), perhaps as a result of the swelling of dendritic buttons or processes [90, 336], but this does not tell us how the patterns of activity are organized. Such improved connexions must, however, involve both excitatory and inhibitory synapses. Certainly, all behaviour involves, not only excitation and inhibition as these terms are used in describing behaviour, but also the excitation and inhibition of neurons. This balance between two opposite effects in the CNS is now, as far as the reflex activity of the spinal cord is concerned, a part of classical physiology: a spinal reflex requires the firing of inhibitory impulses which reduce the contractions of the muscles antagonistic to those producing the reflex movement. Something is even known of the physical basis of inhibition: just as the firing of a neuron by the arrival of impulses involves the depolarization of its membrane, so inhibition probably depends on hyperpolarization [90].

The formidable character of the problems presented by the brain is illustrated by an electrical technique that can be used to study the cortex. This is strychnine neuronography. Small fragments of filter paper, soaked in a solution of strychnine, are applied to points on the cortex in a living animal. The strychnine causes synchronized discharge of the neurons it affects, and the synchrony appears wherever the axons of the strychninized cell bodies are distributed: in all such areas a cathode-ray oscillograph can pick up a 'strychnine spike'. The application of this method to the visual areas of the cortex has been well reviewed by Hebb [133]. An example is given in figures 66, 67: these show Brodmann's areas 17, 18 and 19, which are structural divisions of the main visual cortex in the occipital region, and areas 8 and 20, which have major connexions with the first group. For our present purpose the important facts concern the ways in which groups of neurons in one

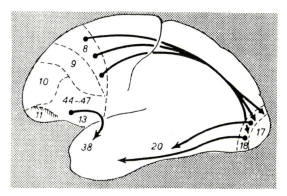

66. Some of the cortical areas concerned especially with vision in man. (After Le Gros Clark [69].)

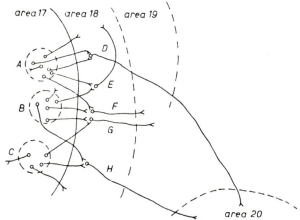

67. Some of the kinds of connexion between cortical areas especially concerned with vision. (After Hebb [134].)

area are connected with the neurons of other areas. The two areas 17 (left and right) receive the optic input from the thalamus, and there is a point-by-point representation of the two retinae in them. Strychninization shows that fibres pass from 17 to 18, but in a diffuse manner: local excitation of 17 is conducted to a large part of 18, and thus topological reproduction of retinal activity breaks down.

Excitation of any part of area 18 is conducted to all other parts of area 18, to the nearest part of area 17, and to all parts of areas 19 and 20. Moreover, there is convergence as well as spread of excitation: for example, cells in different parts of 18 may have connexions with the same points in 20. It is clear that the classical 'reflex-arc' kind of picture of central nervous function will not do for a system arranged like this. What picture *will* serve has still to be discovered.

Other studies of the electrical activity of the brain do not help. We know from the extensive studies of the electroencephalogram (EEG) that the relatively unstimulated brain, or at least the cerebrum, maintains regular changes in electrical potential that can be picked up by electrodes fastened to the scalp. Its rather spectacular character, and its use in the diagnosis of tumours, epilepsy and other conditions in human beings, have led to the EEG being given much attention. It is evidently general in the mammals; Bergen [45] has studied it in the rat. It is, however, not known how the various rhythms of electrical change are maintained or what functions they serve.

The limitations of these methods have led to a search for new ways of studying brain function. Of these the most important are probably biochemical. Much is known of enzymes, vitamins, hormones and other key substances that act on or in the brain. The active substances secreted by neurons, at the synapse or the nerve-muscle junction, especially acetylcholine and noradrenalin, form a large subject on their own; so does the general topic of neurosecretion. But knowledge of brain chemistry is still fragmentary, and the attempt to relate it to behaviour is only just beginning. Some examples are given below, in § 9.6.

9.3 ISOCORTEX AND THALAMUS

9.3.1 *The main structures*

The cerebral hemisphere of a mammal consists of a phylogenetically ancient part which is medial and ventral, and a new part which is much the larger and has until recently been much more studied. The terminology of these structures is confused. Here, that of Pribram [in 125] is used: the 'new', dorsolateral cortex is called the *isocortex*, corresponding roughly to the neocortex or neopallium of comparative anatomy; it is distinguished histologically by having six layers of cells at some stage in its development.

The isocortex may be subdivided according to the character of the thalamo-cortical tracts reaching it. The thalamic nuclei are of two kinds: (i) the extrinsic nuclei receive large afferent tracts from outside the brain; (ii) the intrinsic nuclei receive fibres mainly from other parts of the thalamus. Accordingly there is the extrinsic cortex, corresponding to what we have already called the primary projection areas; and the intrinsic cortex (equivalent to the 'association' areas), of which the various regions receive fibres only from the intrinsic thalamus. This description ignores the motor cortex; however, the latter does in fact receive extrinsic projections from the thalamus; further, the pyramidal (that is, cortico-spinal) tracts originate partly in regions of the cortex other than the primary motor area.

The thalamo-cortical tracts are paralleled by similarly vast projections in the opposite direction. Large cortico-thalamic connexions provide a means by which excitation can pass back to the thalamus as well as from it. Hence there is a structural basis for continual interaction, or 'reverberation', between telencephalon and diencephalon.

9.3.2 Learning and localization

A dominating figure in isocortical etho-physiology for more than a quarter of a century was K. S. Lashley (1890–1958). His work is an outstanding example of an attempt to relate brain function to behaviour. The animals he principally used were laboratory rats, but he also worked on monkeys. The behaviour studied was the learning of discriminations and simple manipulations, and the formation of maze habits. His method was usually to destroy portions of the isocortex, or to make cuts in it, with an electro-cautery, and to study the effects on behaviour. Lashley himself published a late review of his work in 1950 [189]; other valuable discussions are those of Osgood [in 313] and Zangwill [239]; but Lashley's own earlier work should be consulted for its historical importance and the elegance and interest of the writing: much of it has been collected in a single volume [39].

Lashley's most remarkable experiments are those which suggest (i) that learning any task, regardless of its nature and the senses involved, is a function of the whole cortex; (ii) that the ability both to learn new tasks and to retain old skills is proportional to the *amount*, regardless of region, of cortical tissue present ('mass action'); and (iii) that one part of the cortex can take over the functions of others that have been destroyed ('equipotentiality').

One of the concepts strongly criticized by Lashley was that of the reflex arc, with its overtones of a telephone system, as applied to the brain. He made incisions between the visual and motor areas of the cortex of adult rats. He then trained them to make a difficult discrimination: this involved their avoiding X and approaching Δ when these

were presented on a black background, but approaching X and avoiding
Δ on a striped background. This seemed to dispose of any simple
sensori-motor connexions as the basis of learning. However, since this
work was done the extent of the connexions of the cortex with the
thalamus and reticular formation (§ 9.5) have come to be more fully
understood; and these connexions were left intact in the experiments.

More important work involved the destruction of parts of the cortex.
Figure 68 gives an example. Rats were operated on and then tested
for maze-learning ability. Three mazes were used, with one, three and
eight blind alleys respectively. The decrement in learning ability was

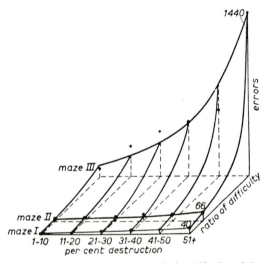

68. Errors in maze-learning increase both with the difficulty of the maze and proportionally to the amount of destruction of the cerebral cortex. (From Lashley
[185].)

proportional to the *amount* of cortex lost, regardless of the locus of the
damage. Rats which had learned to run mazes before operation were
tested for retention of the habit: retention too was found to be proportional to the amount of tissue remaining. Comparable observations
were made on rats which had to learn to release themselves from a
problem box by depressing a platform. In simple learning of this sort,
however, the injured animals sometimes averaged better than the
controls: the latter tended to be too active. In this, as in other instances,
much depends on the character of the problem.

These observations seem to contradict the well-established facts of
cortical localization described in § 9.2.2 above. Yet some other results
go further, since they suggest that the primary projection areas play a

part in learning even when the sense organs which supply the input to them have been destroyed. Rats were, for instance, blinded and required to learn a maze. After this the visual cortex was injured; as before, the loss of habit was proportional to the extent of the damage. Later Tsang reported that rats blind from birth were similarly affected by injury to the visual cortex [321] and this has been confirmed by Lansdell [183]. However, the work of Lashley and of Tsang has been criticized, notably by Finley, on the grounds that the lesions may have extended beyond the visual region [96]; but subsequent experiments by Lashley went some way to meeting her objections.

Other work by Lashley, on visual discrimination, leaves a more confused picture. He first produced evidence that rats' learned responses to different degrees of brightness are impaired by lesions in the occipital third of the cerebral hemispheres; and that the extent of the lesion, as usual, determines the degree of the loss. Later, however, he identified a small region of the visual cortex which is crucial for the learning and retention of visual discriminations: it is injury to this fragment of tissue which, evidently, determines the decrement in behaviour. This, of course, does not fit into the notions of mass action and equipotentiality.

Other observations, too, conflict with an unadorned mass-action hypothesis. Lansdell studied the effects of brain damage on learning in rats which had been regularly handled and had been reared in a complex environment: they were consequently more 'intelligent' than rats kept in the usual small cages (§ 7.4). They were tested by the Hebb-Williams method (described on page 153). Lesions in the front part of the cerebral hemispheres had no effect on performance. Lansdell comments that the anterior cortex, in rats as in human beings, is evidently not important in the solving of simple problems of a kind already familiar to the animal. Performance was, however, substantially affected by injuries in other regions, whether they were made in infancy or in the adult animal [183].

All these observations are of rats. Other work has been principally on primates. Here the evidence for localization is stronger, that for mass action, weaker [239]. This applies to the vast but unsystematic information we have on the human brain, but also to the results of laboratory experiments on monkeys (particularly *Macaca*) and on chimpanzees (*Pan*) [107].

One aspect of the localisation found in the primate isocortex is the existence of suppressor bands or regions (figure 69). If these parts of the cortex are stimulated electrically the result is inhibition of movement on the opposite side of the body. Muscles which are contracted at the

time of stimulation relax and it becomes more difficult to evoke movements by stimulation of the motor area: that is, the threshold of response
is raised. Removal of a suppressor area causes spastic paralysis – persistent contraction of the muscles affected. There is evidently some form
of localization of inhibitory processes. This probably holds for primates
generally, including man.

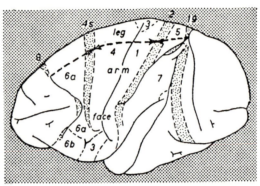

69. Suppressor bands (shaded) in the cerebral cortex of *Macaca*: view of left
side. Area 4 is the primary motor area. (After McCulloch [202].)

More significant for our present purposes is the discovery of secondary motor areas, for instance in the cat (*Felis*) and *Macaca* (figure 70).
These areas resemble the primary ones in having the various regions of
the body represented in them in an orderly way. Injury to the secondary
motor (premotor) area in a monkey on one side causes temporary
spasticity and loss of function; but after a few weeks little deficiency

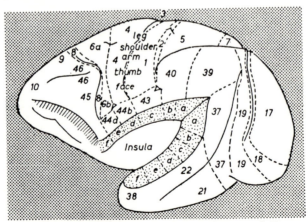

70. Secondary motor areas (shaded) in the cerebral cortex of *Macaca*: diagram
of left side. These areas are actually buried in the fissures of the intact brain.
(After French, Sugar & Chusid [105].)

remains. If the area is destroyed on both sides the animal is much more severely affected, but even then substantial recovery occurs after four to six weeks.

9.3.3 *Studies of 'drive'*

The isocortex is usually thought of as a vast organ of learning; but in fact it also influences both the variability of behaviour and the intensity with which fixed patterns are performed.

Krechevsky studied the effects of isocortical damage on exploratory behaviour in rats. Rats tend to vary their path to a goal when there are alternative paths of similar length (§ 2.2.4). This variation in behaviour was reduced by the lesions; the reduction was proportional to the extent, and independent of the locus, of the injury. In other experiments rats with damaged brains chose a stereotyped path to a goal whereas intact rats preferred a route which was varied by the experimenter [181]. This interesting work was published in 1937. In view of the importance of exploratory behaviour, in the young animal for deutero-learning (§ 7.4) as well as in the adult, it is surprising that it has not been followed up by further detailed studies.

However, Beach has made analogous observations on reproductive behaviour in rats with brain lesions. Maternal behaviour after cortical lesions was impaired in proportion to the size of the injury. In males there was a positive correlation between the amount of cortical tissue remaining and the proportion of animals still able to copulate (figure 71).

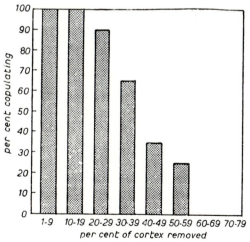

71. Effect of cortical destruction on mating behaviour. The percentage of rats copulating after cortical injury is proportional to the amount of cortex that remained. (From Beach [35].)

A curious feature of this work was that the behavioural deficits could be made up by injecting the appropriate hormone [35, 36].

Much of the work quoted in this and the preceding section suggests that trial-and-error behaviour, exploration and also fixed action patterns depend in certain respects on a general action of the whole of the iso-cortex, at least in so lowly a mammal as the rat. The contradiction of the facts of localization may be only apparent: regions of the cortex which have specialized functions may have a 'mass' function as well; perhaps there is, as Lashley suggested, a general facilitatory influence of all regions. In more complex brains, especially those of the primates, the cortex evidently becomes more differentiated; but even in them there may well be some mass action, perhaps especially during early learning. If there is a general facilitation, it may be asked what connexion this has with the generalized arousal effect, mediated by the reticular acti-vating system, which we discuss below (§ 9.5). It seems that the obser-vations on the rôle played by the cortex in 'drive phenomena' and in learning should be examined in relation to the phenomena of arousal. At present, however, the facts do not justify even tentative conclusions.

9.4 THE LIMBIC SYSTEM

While the isocortex is usually thought of as concerned with learning, the remainder of the cortex, together with certain associated structures, is often supposed to consist of centres of the emotions or sources of drive. For comparative anatomists this system consists mainly of the phylo-genetically ancient archicortex and palaeocortex. Here they are called

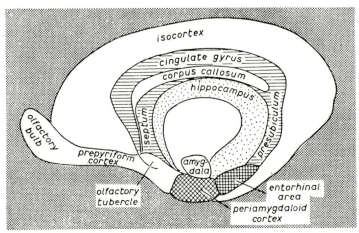

72. Diagram of sagittal section of primate brain to show the main components of the limbic system.

the limbic system (figure 72) and described according to the system out-lined by Brady [in 125]. This system includes (*a*) the receiving centres of the olfactory input and (*b*) a number of other structures with no special association with the olfactory sense. The knowledge we have on their functions has been described as contradictory and bewildering.

The fibres from the olfactory organs first make synaptic contact in the olfactory bulb (figure 13, page 35). Of the neurons in the bulb the mitral cells send afferents to the pyriform lobe (palaeocortex) which forms the lower lateral part of each hemisphere; from there further fibres pass to other cortical structures. Bilateral injury to the pyriform lobe may lead to loss of learned responses to odours. Other (tufted) cells of the olfactory bulb send afferents to the amygdaloid nuclei in the floor of the forebrain. Electrical stimulation of the amygdalae and of neighbouring parts of the cortex (prepyriform) evoke actions used in eating, such as lip movements, sniffing, chewing and licking; salivation also occurs.

Other parts of the limbic system too have been supposed to be con-cerned with smelling; hence the system used to be called the rhinen-cephalon but this term is no longer used. In mammals most of the limbic system, like the isocortex, is freed from attachment to any one sensory modality.

Recent studies of these enigmatic structures have been reviewed by Brady [in 125]. He uses the terms 'palaeocortex' and 'allocortex' synony-mously: they refer only to structures which (i) 'have a clear phylo-genetic primacy' and (ii) meet the criteria for cortical regions of at least three layers of which the superficial layer consists of fibres. This group includes only the hippocampus, the pyriform lobe and the olfactory bulb and tubercle. The term 'juxtallocortex' is applied to cortical regions intermediate between the ancient allocortex and the young isocortex (neocortex): these include the cingulate gyrus and the presubiculum. There is a third group of structures, functionally (it seems) closely related to the first two, but not meeting the criteria for cortex: these include the amygdaloid nuclei and the septal region, both of which seem to have intimate links with the hypothalamus; the caudate nucleus is also tentatively put in this group.

The statement that the limbic system is especially concerned with 'drive' or 'emotion' is based largely on the effects of bilateral injuries, but partly on the results of local stimulation with implanted electrodes. Stamm, for instance, made small lesions in the cingulate cortex or the hippocampus of rats and so produced severe disturbances of maternal behaviour, hoarding and other stereotyped activities [301]. Peretz has shown that such injuries also influence learning: he too made lesions in the cingulate cortex of rats and found that one consequence was a slowing in the rate at which the animals learned to avoid a noxious stimulus [241].

Many workers have reported that damage to parts of the system make the animal fiercer. It has long been known that a mammal with its whole forebrain destroyed can be kept alive for some time: such a 'midbrain' animal is liable to display sham rage, that is, it makes responses normally evoked by a predator or rival in the absence of any appropriate stimulus. It is rather disconcerting, but typical of the present state of our knowledge, that such rages can be induced (in rats and in other species) by injury to the olfactory part of the limbic system. Release of such 'expression of the emotions' can also result from lesions in the cingulate gyrus or the amygdaloid nuclei. These facts suggest that the juxtallocortex and associated structures exert an inhibitory effect on the midbrain system responsible for sham rage. This notion is supported by some results of removing the whole neocortex in cats, but leaving the limbic system: the animals are notably placid and unresponsive to disturbing stimuli.

However, other results quite fail to fit in with this picture. Although, as we saw, injuries involving the amygdaloid nuclei have led to increased responsiveness and sham rage, much work has been done with opposite results. In several species bilateral amygdalectomy has produced a taming effect. Some of the work on rats has been carried out in two stages. First, lesions were made in the septal region (figure 73): these led to wildness of behaviour and also, in some instances, to loss of fear responses which had been induced by training before the operation. Second, in a few animals made fierce in this way, injury to the amygdaloid nuclei has resulted in taming.

The hippocampus, too, is evidently involved in the violent expression of emotion, but its rôle is exceptionally obscure: it has been said that its function changes with each new experiment – a clear indication that the right questions have not yet been asked about it. It has been studied in some detail in cats and monkeys, and also in rats. Stimulation can be carried out with an electrode arranged so that the animal has freedom of movement. When the current is passed the animal gives an impression of increased alertness or, sometimes, of more vigorous arousal expressed as 'anxiety' or even rage. Similar effects are produced by the direct application of cholinergic drugs. Removal on both sides in monkeys leads to increased exploratory behaviour and a reduction in the usual responses to frightening objects; there is also hyperphagia and, in males, increased sexual activity. In the rat hippocampal lesions can lead to loss of fear responses learned before the operation. The hippocampus has two-way connexions with the frontal cortex, but the significance of this is not known. Perhaps it has something to do with the influence of 'drive' on learning. A curious observation, made on the rabbit, is that when there is a fast electrical discharge in the isocortex there is a slow one in the hippocampus, and conversely.

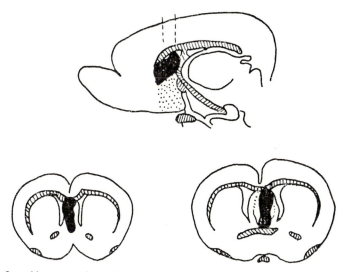

73. Septal lesions in the rat's brain: above, sagittal section; below, two transverse sections. Lesions of this sort lead to wild behaviour in previously tame animals. (After Brady [in 125].)

The nature of 'drive' and the rôle of reward in learning has been obscured, rather than illuminated, by the remarkable work of Olds and his colleagues, already mentioned on page 175. They implanted electrodes permanently in the brains of rats, usually in some part of the limbic system; flexible wires could be plugged in to them when required, and connected so that whenever the rat pressed a lever it gave itself a minute electric shock in the portion of the brain affected. If *not* stimulated, rats in the conditions of these experiments spent 4–10 per cent of their time pressing the lever (according to an arbitrary way of expressing the time spent on different activities). In some positions the electrodes gave scores of 0–1 per cent: stimulation here had a punishing effect. But when the electrodes were in other positions the rats repeatedly stimulated themselves in this way: the stimulation sometimes went on for hours at the rate of thirty responses a minute. The same effect has been observed in cats (*Felis*) and monkeys (*Macacca*). Scores of over fifty per cent were usually observed when the electrodes were in the limbic system; but they also occurred when the anterior (parasympathetic) part of the hypothalamus was affected. Subjectively speaking, the rats and other animals evidently liked the stimulation. There are other observations which seem to support this notion [46], including reports by people who have been stimulated, while conscious, in or near the septal region.

However, cats will not only learn to turn the current on but, if the current then remains on, will learn the way to another point where it can be turned off. Perhaps it is the onset of the excitation that is sought, just as there is a tendency to seek conditions in which there are *changes* in the stimulation of the exteroceptors (§ 2.2.4). A further complication is that, after long periods of self-stimulation, rats become wild and savage; and in the intervals between such periods the animals show a good deal of agitation.

Nevertheless, the high self-stimulation scores may, as Olds [in 125] suggests, occur when the stimulus affects regions especially concerned with the satisfaction of homeostatic needs: normal stimulation of such areas would, perhaps, occur when such a need is met, and would lead to the ending of the activity provoked by the need. (The effect of satiation has been discussed in § 3.4.) Such a system must of course be associated with the parts of the CNS which control overt behaviour, since the latter too has to be adjusted to need.

9.5　THE RETICULAR SYSTEM

The limbic system, though baffling, is evidently concerned in some special way with 'drive'. The same applies to the reticular formation or substance of the brain stem; indeed, some of the limbic structures may be regarded as specialized anterior portions of the reticular substance. In most elementary accounts of the brain, however, the brain stem is described as an arrangement of well-defined reflex centres. These centres are highly specialized in function and can operate independently of the forebrain: a 'midbrain animal', in which the forebrain has been destroyed, still has an array of reflex responses. It is therefore a paradox that the midbrain also has a component of which the functions are generalized and not specific and which operates in intimate relation to the forebrain. This component, the reticular activating system (RAS), consists of large numbers of small cells, most of them without long axons, whose functions were for long an enigma. It was only in 1949 that Moruzzi & Magoun published observations which have led to a great outburst of researches and which have given this tissue a prominent position in neurophysiology. The subject has been reviewed by Jasper and colleagues [162] and by Magoun himself [206].

The reticular substance is a matrix or network of cells extending from behind the medulla into the thalamus. It receives collateral fibres from afferent tracts as they proceed towards the thalamus, and all the receptors contribute inputs to it, directly or indirectly: hence it is a centre, or group of centres, on which information from every sensory modality converges. There are also cortico-reticular tracts, both from the isocortex and the limbic system. As for outgoing connexions, it has long been known that there is a substantial diffuse projection system of fibres from the reticular substance to the whole of the cortex; this is in contrast to the more familiar specific projection system carrying the sensory inputs with only one synaptic break in the thalamus, to the special sensory regions.

Although the RAS is often spoken of as a single entity, in fact it can be divided into (i) a posterior part in the medulla and cervical cord, (ii) a central region and (iii) a thalamic reticular system.

Most attention has been paid so far to the central part, which lies largely in the midbrain; it is to this part that injury causes coma. Consider a sleeping animal of which the EEG is being recorded and in which an electrode has been implanted so that the reticular substance can be stimulated. When the current is switched on the EEG alters to that of an awakened animal, and the animal may indeed wake up and become responsive to its surroundings. The recording of impulses in various parts of the brain shows that a sleeping or anaesthetized animal still receives the input in the specific afferent tracts from the extero-ceptors, but that the diffuse projection system is silent: evidently responsiveness depends on the additional cortical facilitation provoked by impulses from the reticular system. This, as we saw in § 8.3, is the basis of Hebb's identification of the RAS as the source of 'general drive'.

As might be expected, the reticular system, though called 'activating', is not only stimulatory in function: its influence can be inhibitory; this applies especially to the anterior (thalamic) and posterior (medullary) regions. Stimulation of the thalamic portion brings about the 'arrest reaction', that is, immobility such as is often seen in a frightened animal. This suggests that the thalamic RAS is associated in function with the nearby limbic centres which, too, are concerned with the expression of fear and anxiety.

The inhibitory effect of the posterior portion is of a different kind. The reticular formation does cause generalized arousal, but its action is not as indiscriminate as this statement suggests. Work on cats (*Felis*) has provided good examples of the fact that the system is concerned with *selection* from the great volume of information flooding in from the sense organs. For instance, if a cat, awake but sitting still, is subjected to a sound such as a regular clicking, impulses can be duly recorded in

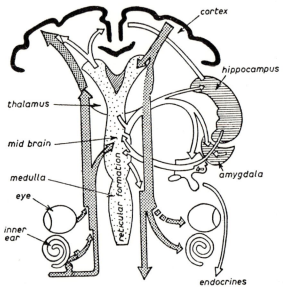

74. Some of the connexions of the limbic and reticular systems.

the cochlear nerve; these impulses are relayed to the isocortex. If, however, in this situation, the animal is then allowed to see some mice, or to smell fish, the auditory input is reduced: it is as if the animal becomes partly deaf to these 'neutral' or 'meaningless' sounds when presented with a significant stimulus. That this kind of effect is mediated by the reticular system is inferred from inhibitory effects observed when the posterior region is electrically stimulated. The inhibitory influence of the reticular formation is quite general. There are even afferent tracts from it to the periphery, so that the sensory input can be influenced before it reaches the brain. This recalls the *Reafferenzprinzip* of von Holst & Mittelstaedt, mentioned above on page 214.

We have seen, in earlier chapters, that the selection of what features of the environment to respond to, and what to ignore, often depends on learning: all complex learning includes habituation as a component

(§ 6.2.2). The reticular formation is concerned in this aspect of discrimination also. In general, a repeated stimulus loses its effect on the
RAS, just as it does on overt behaviour. If the stimulus to which the
animal is habituated is a tone of given frequency, the sounding of
another frequency produces a full response in the RAS; so does a
stimulus, such as a puff of air, which affects a different sensory modality.
Sharpless & Jasper have produced similar habituation to a simple
melody or 'pattern of several tones'; when this is established, sounding
an isolated tone from among those that form part of the pattern causes
activation of the RAS. Here too, then, is a neural counterpart of a
familiar behavioural phenomenon [285]. This observation illustrates
the complexity of function in which the RAS is involved.

It is obvious that the RAS plays an essential part in the phenomena
of arousal, 'general drive' (§ 8.3) and habituation. Nevertheless, the
problem of 'localization of function' arises with the reticular formation
as it does with other structures. Sharpless & Jasper found that habituation to single tones occurred in cats in which the whole of the auditory
cortex had been removed on both sides; but differential habituation to
melodies required an intact cerebral cortex [285]: this degree of complexity of learning evidently depends on the reticular formation and
cortex acting in combination.

A similar interrelationship can be demonstrated during the establishment of a CR. This has a special interest, since Pavlov's classical work
was explicitly designed to throw light on the 'highest' part of the brain,
the cerebral cortex. Magoun [206] and Gastaut [in 162] quote work in
which the electrical activity of subcortical structures was recorded
during establishment of a CR. The unconditional or releasing stimulus,
in some of the experiments, was a flickering light: this induced a neural
discharge of a specific frequency. A sound was used as a conditional
stimulus: after training, the sound alone evoked the pattern of discharge
previously due to the light. This conditional discharge appeared in the
midbrain reticular formation before it was evident in the cortex; it was
also stronger in the RAS, and longer-lasting there on each occasion on
which it was evoked.

These observations give depth to the general statement that the RAS
provides a facilitatory effect in all behaviour. They also illustrate once
again the way in which the cerebral cortex continually interacts with
other structures (figure 74). They have indeed led some authors to put
special emphasis on connexions between the cortex and the more
posterior structures and to play down those cortico-cortical connexions
which had previously been thought to be all-important. But the fact

that there are continually active circuits involving both cortex and brain stem or thalamus does not imply that communication within the cortex is of no significance. The work on CRs quoted above was successful because a very simple form of learning was studied – one in which the motivational aspect could be ignored. There was consequently a simple electrical pattern in the brain which could be readily recorded. In more complex learning, for instance that involved in discriminations, and in learning accompanied by a higher level of arousal, the pattern would doubtless be more difficult to unravel.

The interaction between cortex and RAS is of course a two-way process. Cortico-reticular tracts have already been mentioned. French [102] has discussed work on them in the monkey (*Macaca*). Only some regions of the cortex give origin to these fibres; if one of these regions is stimulated the result is a desynchronization of the EEG of the kind seen when the RAS itself is stimulated.

9.6 METABOLISM

There remain to be described some pioneering studies of the relationship between brain chemistry and behaviour. This field is almost unexplored, but at least one group of researches exists to suggest that it is already worth exploring.

A number of substances are known, from observations of behaviour, to have specific actions on the central nervous system. The substances include carbon dioxide, various hormones and other constituents of the blood. Sometimes, as with thiamin (vitamin B_1) something is known of the cellular chemistry of their actions. Among the important substances acetylcholine (ACh), the neurohumour of neuromuscular contraction, occupies a special place, since it is probably the chemical agent also of synaptic transmission in the CNS. Wherever ACh is important, so is cholinesterase (ChE), since this enzyme performs the essential rôle of destroying ACh rapidly as soon as it is liberated.

Krech, Rosenzweig, Bennett and their colleagues have studied ChE activity in the brains of rats, in relation to features of behaviour which differ in different strains. In some of their experiments the rats were set a maze problem which could not be solved: the pattern of alleys was altered in a random manner between runs. At each choice point the animal could turn left or right; it could also go either into a lighted or a dark alley. Whether the light alley was on the left or the right was varied at random. Neither the spatial cue nor the visual one led reliably towards the goal: reward was consequently sometimes attained, but unpredictably. In these conditions rats tend to choose *either* spatial 'hypotheses' (such as turning regularly to the left), *or* visual 'hypotheses' (such as turning regularly towards the dark). The rats which tended to make spatial, rather than visual, hypotheses had the greater ChE activity in the isocortex. This was not a matter of a general difference in enzyme activity, since at least one other enzyme, lactic dehydrogenase, showed no such correlation with behavioural traits.

Further study, in which rats were given problems which could be solved, showed that those with the higher ChE activity were the more adaptable: they varied their hypotheses more readily. At this point it seemed that high ChE activity was a concomitant of 'intelligence', in the sense in which that term has been used in § 6.4. ChE was also considered to be an index of ACh activity, and so of readier synaptic transmission, but this notion is perhaps rather speculative. In any event, study of other strains of rats showed that the correlation between ChE and learning ability does not hold for all strains. At present the possibility that the relationship between ChE and ACh is a more complex one is being studied: it is still not ruled out that the acetylcholine level is an important determinant of intelligence [263].

The same workers have also published observations in which ChE activity was the dependent variable: that is, they have asked whether environmentally imposed differences in behaviour influence brain ChE. They compared the effect of rearing rats in a complex environment with that of being reared in a small cage; the restricted animals were themselves in two classes: some were in groups of three to each cage, while the others were alone. As described in § 7.4, the rats of the first group developed a superior learning ability; and these rats were found to have the lowest ChE activity in the 'dorsal cortex' (that is, in effect, the isocortex), and the highest ChE in the rest of the brain. Those reared alone had the highest isocortical ChE and the lowest subcortical ChE, while the members of groups of three were intermediate [178]. A curious additional finding was that unilateral cortical lesions in rats are followed by increased ChE activity on the intact side [177].

It is difficult to know what to make of these results. The authors themselves do not attempt any firm interpretation, since they consider that this would be premature. The need for caution is strengthened by the studies of Hydén [159] on the chemistry of neurons and of glial cells. The latter are non-conducting cells present among neurons in all parts of the brain: they are some ten times as numerous as nerve cells, occupy about the same volume but are chemically quite different;

they consequently add greatly to the difficulties of discovering anything useful about the chemistry of neural function.

9.7 CONCLUSIONS

It has been possible to keep this chapter short because, despite the vast amount of work on the mammalian brain, we know exceedingly little about the way the brain organizes behaviour. Nevertheless, even if we are at present gravelled for lack of matter, we must not reject study of the brain as a source of knowledge about behaviour. To do so would be analogous to refusing to study the kidney when investigating the composition of the urine.

The brain, however, presents more difficulties than the kidney. Elementary accounts of the nervous system tend to conceal this fact. For one thing, they often speak of the 'functions' of parts of the brain in a misleading way. Three of the preceding sections are headed respectively 'cortex', 'limbic system' and 'reticular system', but this anatomical arrangement does not correspond to the facts of function: the study of any one of these systems soon becomes meaningless without reference to the others. During every few milliseconds, in the waking brain, information passes to and fro in a network of communication of which only the larger details are yet certainly known. In every specious present the picture of the outer world which reaches the CNS from the senses is subject to a system of editing or selection and is also imposing changes, more or less lasting, which affect subsequent behaviour. In such a flux we cannot, with our present knowledge, properly speak of localization of function, but only of the specific effects of injury or stimulation: the fact that, say, the expression of fear is altered by damage to a single structure by no means shows that control of that behaviour pattern is localized there: it tells us only that the structure is necessary for the normal performance of the behaviour. A small injury can influence behaviour which certainly depends also on the functioning of other parts; by contrast, some substantial injuries leave behaviour largely unaltered; and, when behaviour is disturbed by lesions, there may be subsequent recovery due, evidently, to some compensatory process elsewhere.

These facts at present defy explanation. All they do is to make accounts of neural function in terms of reflex arcs as absurd as interpretations of learning in terms of CRs.

The brilliant work of a few men, in this field of formidable problems, now enables us to define some of the difficulties with precision. It also gives students of behaviour a picture of a functioning entity which can help them to avoid excessive resort to what Isaac Newton called 'occult qualities', such as 'instinct', 'drive' and even 'learning'. Instead of thinking of the animal as a black box with a totally obscured interior we can at least dimly see how the complexities of overt behaviour may eventually be illuminated by nerve physiology.

The Study of Behaviour

I do not know what I may appear to the world; but to myself I seem to have been only like a boy, playing on the sea shore, and diverting myself, in now and then finding a smoother pebble or a prettier shell than ordinary, whilst the great ocean of truth lay all undiscovered before me.

ISAAC NEWTON

This book is primarily about the principles of the study of behaviour, sought and propounded for their own sake. The detached investigation of nature, motivated by disinterested curiosity, is often accepted as a worthy pursuit. So it should be. But it is compatible with an awareness of the social rôle of scientists and their researches. The behaviour of *Rattus* (the genus from which most of the examples in this book are taken) is often studied for its possible significance in other fields. Much of the work on laboratory rats is partly motivated and endowed for its putative value to human psychology. Knowledge of exploratory behaviour, or of the effects of reward and punishment in learning, could conceivably have implications for the upbringing and education of children. Study of abnormal behaviour might contribute to understanding of human psychopathology. More certainly, in another area of applied biology, knowledge of the behaviour of wild rats has helped to prevent disease and to preserve food on farms and in warehouses.

Occasionally, in this book, mention has been made of such implications. There is everything to be said for presenting the special results of academic inquiry, not only against a general scientific background but also in relation to their possible uses. If this is done effectively two ends may be achieved: to introduce a reader to some fragment of the strangeness and complexity of the natural order; and to show what use may be made of the little knowledge so far gathered. In the middle of the twentieth century we are faced with formidable problems of controlling both our own behaviour and also our relations with other species. To solve them we shall have to use every available fragment of knowledge and wisdom.

Whether one is concerned with the applications of ethology or not, one is obliged, while actually investigating animal behaviour, to adopt a detached attitude. Whatever use may eventually be made of one's results, the observations themselves, and the theories proposed, should be free from bias. In § 1.4.1 some examples were given of the effects of unconscious preconceptions. We can perhaps now see more clearly how bias or 'set' can influence, not only the interpretation of what one observes, but also the pattern of research or teaching. For many years the exceedingly obvious exploratory propensities of rats were almost ignored: most of the workers on rat behaviour were studying learning, often within a rather rigid theoretical framework: exploratory activity seemed irrelevant or inconvenient – when it was noticed at all. After all, the century had begun with Pavlov's work, in which the experimental conditions precluded all exploratory or appetitive behaviour. Bertrand Russell once suggested that all careful observations of animal behaviour confirm the philosophy of the observer and reflect his national character.

> Animals studied by Americans rush about frantically, . . . and at last achieve the desired result by chance. Animals observed by Germans sit still and think, and evolve the solution out of their inner consciousness . . . the type of problem which a man naturally sets to an animal depends upon his own philosophy, and this probably accounts for the differences in the results. [264]

Since 1927, when this rather frivolous passage was published, the highly acquisitive and competitive society of North America has seemed to provide examples which support it: there has been a remarkable amount of work on hoarding by rats; and studies of social behaviour have displayed an excessive preoccupation with relationships of dominance and subordination. Meanwhile, the study of 'instinct' has flowered in the conservative lands of Western Europe; and this study at first involved a neglect of ontogeny and an unwarranted assumption of rigidity in behaviour.

A reassuring feature of the history of ethology is that all these different attitudes to research have been fruitful. Certainly, once a topic for study has been chosen a high standard of rigour in the design of experiments and the interpretation of results has often been achieved. In any case, the divisions of outlook and method between the different schools are now disappearing. It is this which has made writing this book possible.

The attempt to achieve rigour has led to that meticulous avoidance of subjective modes of expression which makes some writing about animal behaviour read rather oddly. (No doubt this book provides many examples.) It perhaps does not matter very much whether one refers to a 'hungry rat' or a 'rat which has fasted for twelve hours'; the second phrase is of course more informative, since it is quantitative: without the last three words it is equivalent to the first phrase. More important than objective phraseology is the fact that we now habitually use a unitary or monistic way of speaking about even our own behaviour. There are several examples of the value of this method in the preceding chapters.

Consider for instance the phenomenon of death due to 'social stress' (§§ 4.3.2; 8.2.3.2). If we were to use the dualistic language colloquially employed for man we should attribute the condition of the stricken rats to 'mental' rather than to 'physical' causes. In conversation people are inclined to talk of 'shock' or 'humiliation'. But these expressions have little or no explanatory value. One of the distant aims of this research was to be able to say that death in the circumstances described in § 4 is due to certain processes in the body; and that it could be prevented if one interfered with those processes. This assumes that the methods of physiology are capable of revealing the causes of this kind of death. At present it is possible to describe only some of the internal changes that accompany social stress, but there are no grounds for assuming any totally new principle or separate mode of being involved in this (or any other) aspect of behaviour.

This does not imply that a unitary interpretation must be physiological. We have already seen, in § 1.4.1, that certain sorts of explanation are expressed in purely behavioural terms. The death of rats under attack again provides an example. It may be asked why attack, without wounding, by a member of the same species has such a violent effect. It is possible to suggest a partial and tentative answer by referring to other work on behaviour. There are certain kinds of situation that cause flight in other species of mammal. Hebb has described how chimpanzees are terrified by the stuffed head, without the body, of another chimpanzee: in general it is frightening to see something which has, like a ghost, only *some* of the attributes of a familiar object [131]. The avoidance of new objects displayed by wild rats (§ 2.3) is a similar phenomenon. It may be suggested that this neophobia has a counterpart in the meeting of a newcomer (an interloper) with a resident male. We saw in § 4.2 that rats are in general amicable creatures: attack, or at least severe attack, is an anomaly in the lives even of wild rats. It represents, perhaps, a sharp contrast with the norm or the 'expected'; and it may be this that induces the intense state of fear observed in the attacked rats [29]. Whether this is valid or not, the explanation proposed is

behavioural and is independent of physiology. It is of course less satisfying than a physiological one (for reasons already discussed in § 1.4.1):
physiology enables us (at least in principle) to deduce the properties of a
system from its own organization; a generality about behaviour may
have predictive value, but no more.

The preceding passage implies that correct prediction is desirable. No
doubt few will quarrel with this assumption. In practice, correct prediction is the test of the validity of many, perhaps most, of the general
propositions asserted by scientists. It has been called a 'metaphysical
directive' – which perhaps means no more than Euclid meant by
'axiom'. Another such directive is usually referred to as Occam's razor
and misquoted as *entia non sunt multiplicanda praeter necessitatem.*
(Occam actually wrote: *pluralitas non est ponenda sine necessitate.*) In
any case, these exhortations express a general refusal to assume the
existence of more, or more complex, entities than the least and simplest
needed to explain the facts. (It might be interesting to make a search for
instances in which adherence to this principle has led to error.)

The techniques of observation, experiment and communication used
in the study of behaviour have been developed in the quest for reliable
methods for answering specific questions rigorously stated. It is rather
rare for a psychologist, for instance, to ask (as a layman may reasonably
ask), 'what is consciousness?' (In the same way, biologists do not often
ask, 'what is life?') They ask, say, what are the effects of injury to the
reticular formation on a particular feature of behaviour [271]. One of
the most important elements in human history of the past three hundred
years has been the growth of methods which enable us to answer such
specific questions with a good deal of confidence.

Of course, ethology has also some general principles to offer, though
these give us only a crude and provisional conceptual framework at
present.

(i) One is the truism that an animal's behaviour contributes to the
fitness (in the strict, Darwinian sense of the term) of its species at least.
(*a*) Some behaviour does so by being closely linked to homeostatic
processes: as a rule, an animal drinks when it needs water, eats when it
needs a source of energy and so on: in this way, osmotic relationships,
body weight, body temperature and other quantities are kept steady.
This entails a complex system of negative feedbacks, operated especially
by the nervous system. (*b*) But reproductive behaviour too contributes
to fitness, though not to homeostasis: that is, it does not influence

steady states on which the animal's life depends (though maternal care is needed for survival of the young of many species). (*c*) Behaviour may contribute to fitness without being directly linked either to homeostatic processes or to reproduction. This holds for exploratory behaviour, which (*a*) gives information which may later be of use (through the accompanying 'latent' learning), (*β*) increases 'intelligence' in young animals (by making deutero-learning possible). It also holds for imprinting and the learning of songs by birds.

(ii) Much behaviour, notably that involved in mating, the care of young, defence of territory and the construction of nests and so on, is highly stereotyped and predictable in detail for each species. We consequently speak of fixed action patterns. Most land animals, and many aquatic ones, have species-characteristic behaviour patterns or other features which evoke standardized responses in other members of their species. These social signals ensure either (*a*) approach and subsequent mating or parent-young interactions; or (*b*), where territory is involved, flight. In the past, behaviour of this sort has been called 'innate' or 'instinctive'; but it is now seen to occupy positions in a continuum of acts of which the development ranges from highly labile to relatively fixed.

(iii) One of the notable features of behaviour in general is its lability: most animals, perhaps all, can alter their responses to a given class of stimuli in accordance with their experience of 'reward' or 'punishment'. The internal processes, no doubt very diverse, which underlie such alterations, are all lumped together under the excessively general term, 'learning'. This term also covers the processes which take place during exploration, imprinting and so on, mentioned in (i) above.

(*a*) The form taken by learned behaviour is often determined by the fact that a particular response satisfies a 'need'; more precisely, there are certain consummatory states, not all homeostatic in the sense given above in (i *b*), which an animal tends to learn to achieve. Such learning commonly takes place during variable movements (called appetitive behaviour) which resemble exploration but are in such a case impelled by a specific internal state such as hunger. (*b*) The goals which an animal tends to learn to achieve are not constant. There are often autonomous changes (as in breeding cycles) in an animal's internal condition which cause corresponding cycles in behaviour. In any case, attainment of a consummatory state is usually followed by quiescence, or by behaviour directed towards another goal. (*c*) In a few instances the 'drives' or internal states which impel an animal towards specific goals have been partly analysed. It is then found that the processes involved are multiple: not all are *directly* homeostatic even when they are concerned with homeostasis; in particular, an activity such as eating or drinking may cease as a result of afferent impulses which are independent of the internal change which initiated eating or food-seeking.

(iv) Learned behaviour, except in its simplest forms, is not a matter of the substitution of one response (or no response) for another: it cannot be described in terms of the classical conditional reflex. It involves (*a*) selection of one set of responses from among many, in trial-and-error behaviour; (*b*) processes of generalization and discrimination; and (*c*) 'insight behaviour' in which separate previous experiences are used in combination to make a pattern fitted to a new situation. Learned behaviour is typically anticipatory. Some is also performed as if a calculation of probabilities of the results of alternative actions had been made.

(v) Behaviour varies in the individual in three ways, of which two have been mentioned above: (*a*) there are the fluctuations of 'drive'; (*b*) there are the adaptive changes due to learning; (*c*) there are developmental changes. As for the last, sometimes a fixed action pattern can be evoked only when the animal is sexually mature; other patterns appear at birth (or hatching or emergence). Neither the time of their appearance, nor their rigidity, tells us anything conclusive about how they develop. Some fixed patterns develop normally only if the young animal is subjected to specific experiences during an early sensitive (critical) period. Sensitive periods are also important in the development of 'intelligence': the deutero-learning mentioned in (i) above can take place fully only in early life.

(vi) All development involves an interaction of (*a*) the coded information in the genes with (*b*) the animal's environment. Some behaviour patterns are not only stereotyped but also exceedingly stable in their development: they seem to appear in virtually any environment which allows the animal to develop at all. Others are more labile: they depend a great deal on the particular sensory inputs experienced by the individual. Some sensory effects, such as those of a nestling in contact with its mother, are inevitable except in very special experimental conditions. Others are variable and are responsible for most learned behaviour.

(vii) The central nervous system connects the sense organs to the effectors and is also an organ of learning. It selects from the input only certain patterns and organizes the motor output either in species-characteristic or in individually learned patterns. We do not know how it does this.

It is appropriate to end this summary of principles with a statement of ignorance. Our knowledge of the facts of behaviour is growing quickly, but we have as yet taken only a minute sample of the varieties of behaviour to be found in the animal kingdom. There are several rigorous methods of study, each fruitful, including those of field ecologists and neuro-physiologists as well as those used in the direct study of behaviour. A few generalizations, valuable but of small scope, can be made, at least on the plane of overt behaviour; but the underlying physiology is baffling, and no grand synthesis is yet in sight. Newton's great ocean of truth remains all undiscovered before us.

Appendix 1: On Definition

It is just because the terms of science are so well defined, and defined in a way which is closely tied down to the phenomena, that questions in science can be settled: only because this is so can scientists hope to answer definitely the questions that arise for them, by looking to see whether things actually happen in nature in the manner the theory suggests.

STEPHEN TOULMIN

In the writing of this book it has been assumed that precise and consistent definition of technical terms helps in conveying what is meant. This may be regarded as an hypothesis about human behaviour. It is, of course, based on the experience of teachers and others. There are successful scientists who dispute, or at least ignore, this view. Some say that words can never match the complexity of events and that semantic rigour is unattainable. The reply to this is that an approximation to rigour is nevertheless possible and useful. Others say that definitions must always be provisional: meanings change as a subject advances. This, however, is not a cogent argument against agreed definitions in a given work.

The word 'definition' has itself several meanings, fully discussed by Robinson [259]. For example, an *ostentive definition* is one in which an object is indicated to an audience and the name to be attached to it is then given. A sophisticated example is the formal naming of a new species by a taxonomist.

In this book two other sorts of definition are more important. The first is *lexical definition*. In a dictionary dealing with current usage the assertions about the 'meanings' of words are historical statements: they purport to tell the reader how words have in fact been used in the recent past (and by implication how they are being used now); they may consequently be true or false. The second is entirely different. It is *stipulative definition*. This is a statement of intention. The writer says, in effect: from now on I shall use this term in this particular way and you, the reader, are asked so to understand it. This is what Humpty Dumpty had in mind when he said that words meant what he intended them to mean. A stipulative definition cannot be true or false, unless one applies these terms to the writer's success or failure in keeping his word.

It is therefore never admissible for a critic to complain that stipulative definitions are 'incorrect'. Admittedly, such a definition may not conform to current usage (as in the chemists' definition of the term 'salt'); and it may for that or other reasons be *inconvenient*. But it can never properly be said to be *wrong*.

The stipulative definitions that one adopts are usually based on an accepted convention. The more this is so, the better, since the result will be the easier to read. But when terms commonly used are confusing it may be necessary to disregard them and to ask, what are the facts?; and then to try to describe the facts in neutral and unambiguous language. This involves a departure from usage which is often very difficult to make.

Sometimes a completely new term, such as 'conditional-reflex', has to be coined; in such a case – given care in the initial defining – no trouble should follow. At other times one may avoid difficulties by discarding a much-used but ambiguous term, such as 'instinct'. When this is done it is still necessary to discuss the facts, concepts or theories subsumed under the rejected term. Precise definition is no substitute for observation, experiment or the formulation of hypotheses.

Appendix 2: Glossary

A lexicographer is, *pace* Dr Johnson, something more than a harmless drudge.
A. J. AYER

The notes which follow are primarily designed to state how the terms listed are used *in this book*. Definitions have been kept short because longer ones (which might achieve greater rigour) are probably not as informative as simple definitions combined with the examples given in the main text. Where an entry is followed by '[not used]' this usually means that the term has been mentioned only in a discussion of facts and concepts to which the term is applied by others. In addition to giving stipulative definitions (see Appendix 1), the entries also constitute *recommendations*, in as much as the author has found the usages convenient. There is room for much improvement still; but it is likely that this will for the most part have to await further advances in knowledge of behaviour and physiology.

For a comprehensive list of terms and meanings, with caustic comments, see Verplanck [323].

ANXIETY

A state, usually aroused by noxious stimulation, in which there is autonomic, especially sympathetic, activation. External signs include more-than-usual defaecation, urination, grooming and other activity. (Fear is the same, but 'fear' is customarily used only if the condition lasts for a short time.) § 8.2.3.

APPETITIVE BEHAVIOUR

Variable behaviour, often impelled by an internal state such as hunger and ended by the abolition of the internal state. See *consummatory state, drive*. (The concept of appetitive behaviour is unsatisfactory because of its vagueness and excessive generality.) §§ 1.3; 2.2; 6.2.3.

CONDITIONAL REFLEX (CR)

An act, elicited by a previously indifferent stimulus (CS), as a result of repeated application of the CS at about the time of application of a releasing stimulus for a similar act. [This is 'classical', 'Pavlovian' conditioning, or 'CR type I'.] The CS is 'indifferent' only with respect to the activity to be studied; it must of course arouse attention from the first. §§ 6.2.1; 7.2.

CONDITIONED REFLEX

See *conditional reflex*.

CONSUMMATORY ACT

A stereotyped activity which ends a complex sequence of behaviour. § 1.3.

CONSUMMATORY STATE

A state towards which behaviour tends to be directed. (For instance, a full belly or a particular skin temperature.) § 1.3.

DEUTERO-LEARNING

Improvement in learning ability due to diversity of experience in early life. § 7.4.

DISPLACEMENT BEHAVIOUR

[not used] § 8.2.

DRIVE

An internal state causing increased activity. When used in this book it is accompanied by an epithet, as in 'hunger drive'; but see also a discussion of the concept of 'general drive' (§ 8.3). § 8.1.

ETHOLOGY

The scientific study of animal behaviour.

EXPECTANCY

A representation in the central nervous system of stimuli which have in the past followed the situation in which the animal now finds itself. § 7.3.2.5.

EXPLORATORY BEHAVIOUR

Any behaviour which tends to increase the rate of change in the stimulation falling on an animal's receptors, and which is not impelled by homeostatic or reproductive need. §§ 2; 8.4.2.

EXPLORATORY LEARNING

Unrewarded learning that becomes evident only on a later occasion when there is some incentive to use it. § 7.3.1.2.

EXTINCTION

Decline in performance of a learned act as a result of repeated evocation of the act without subsequent reward. §§ 6.2.1; 7.3.3.

FEAR

See *anxiety*.

FIXED ACTION PATTERN

Stereotyped, highly predictable, taxon-specific behaviour sequence. [This kind of behaviour is still often called 'innate' or 'instinctive'; but these terms have not been used in this book because of the confusion which surrounds them.] §§ 5.1.1; 5.4.

HABITUATION

Decline in performance of an act as a result of repetition of a stimulus which evokes the act. [Often reserved for decline in a *stereotyped* act such as a reflex.] See *extinction*. § 6.2.2.

IMPRINTING

The learning process involved in developing, during an early *sensitive period* (q.v.), the tendency to follow or otherwise approach an object (usually another animal and in ordinary development one of the animal's own species). § 8.4.1.

INHIBITION

Any internal process which prevents or reduces the performance of an action. [In nerve physiology has a different and more specific meaning.]

INHIBITION, INTERNAL

Synonym for *extinction* (q.v.).

INNATE BEHAVIOUR

[not used] See *fixed action pattern*.

INNATE RELEASING MECHANISM (IRM)

[not used] See §§ 5.1.4; 5.2.1.2.

INSIGHT BEHAVIOUR

A sudden adaptive reorganization of behaviour. See *expectancy, exploratory learning*. §§ 6.3; 7.3.2.5.

INSTINCT

[not used] See § 5.4.

INSTRUMENTAL CONDITIONING

[not used] See *trial-and-error*.

INTELLIGENCE

Learning ability. (Not a useful term, unless the behaviour involved is specified.) § 6.4.

LATENT LEARNING

Synonym of *exploratory learning* (q.v.).

LEARNED BEHAVIOUR

Advantageously changed individual behaviour due to (usually repeated) stimulation. (This term is so vague and general that it is hardly useful in an account designed to be rigorous. The kinds of behaviour to which it refers are very diverse.)

LEARNING

Internal change causing adaptive change in behaviour as a result of experience [311]. See *learned behaviour*.

MOTIVATION

[not used] The fluctuating internal states which determine what kinds of behaviour and what intensity of activity can be evoked at a given time. § 8.1.1.

OPERANT CONDITIONING

[not used] See *trial-and-error behaviour*.

PUNISHMENT

The encounter by an animal with conditions which it tends to avoid. § 7.3.2.4.

REINFORCEMENT

The encounter by an animal with conditions which it tends to seek. § 7.3.2.

RELEASER

[not used] See *releasing stimulus*.

RELEASING STIMULUS

A stimulus (q.v) that evokes a specific (stereotyped) response. [In some writing this term is reserved for stimuli arising from other members of the same species; but these may be conveniently called 'social releasing stimuli'.]

SENSITIVE PERIOD

A period in an animal's life, usually early, when it is especially easy to evoke a particular kind of behaviour or a particular learning process. § 8.4.1.

SPECIFIC ACTION POTENTIAL(ITY)

[not used] See § 8.1.1.

SIGN STIMULUS

[not used] See *releasing stimulus*.

SOCIAL RELEASING STIMULUS

See *releasing stimulus*.

STIMULUS

An event which excites any of an animal's receptors.

TRIAL-AND-ERROR BEHAVIOUR

Gradually learning to solve a problem; the process involves an increase in efficiency brought about by elimination of unnecessary movements. (This is equivalent to, or includes, the categories of 'CR type II', 'operant conditioning' and 'instrumental conditioning'. It differs from 'classical conditioning' – here called CR formation – in that there is an appetitive phase during which at first a great variety of ineffectual movements may be made.) §§ 6.2.3; 7.3.

Acknowledgements

I am grateful to Michael Abercrombie for suggesting that this book should be written. R. A. Hinde very kindly read most of the typescript and made valuable criticisms. Help with portions of the book was generously given by G. C. Kennedy and Joseph Schorstein. I am much indebted to all these colleagues for their advice. Special acknowledgement is due to Gabriel Donald for his drawings of wild rats in action.

Bibliography

RECOMMENDED REVIEWS

General

Munn's *Handbook* [233] gives a comprehensive review with a complete bibliography of rat psychology up to December 1949. Osgood's *Method and Theory* [239] is a valuable source book for experimental psychology in general. Roe & Simpson, *Behavior and Evolution* [260] gives a biological dimension to studies of behaviour. Thorpe's *Learning and Instinct* [312] reviews work of learning in the whole of the animal kingdom. Young's *Life of Mammals* [337] provides a foundation for the study of mammalian biology.

§ 2: movement
  Barnett [21]; Berlyne [46].

§ 3: feeding behaviour
  Soulairac [296].

§ 4: social behaviour
  Barnett [19] for the social behaviour of wild rats in artificial colonies; Hinde [147] for a discussion of territorial behaviour; Marler [209] for a review of animal communication; Tinbergen [315] for an introduction to social behaviour in insects, fish and birds.

§ 5: fixed patterns
  Beach [37] for hormones and behaviour; Fuller & Thompson [106] on genetical variation and behaviour; Larsson [184] on sexual behaviour in male rats; Lehrman [194] on maternal behaviour and 'instinct'.

§§ 6 and 7: 'learning'
  Hebb's *Organization of Behavior* [133] remains stimulating reading; two more recent works by Mowrer [230, 231] not only give much well-argued material but also an historical treatment; Osgood [239]; Thorpe [312].

§ 8: 'motivation' or 'drive'
  Hinde [145, 150].

§ 9: brain
  Bonin [53] and Fulton [107] have both written conveniently short

books, mainly on the cerebral cortex. Harlow & Woolsey [125] have edited a long, impressive, wordy and difficult symposium.

LIST OF REFERENCES AND AUTHOR INDEX
The following abbreviations have been used:

AB   Animal Behaviour
AJP   American Journal of Physiology
BJAB British Journal of Animal Behaviour
CRSB Comptes Rendus de la Société de Biologie de Paris
JCPP Journal of Comparative and Physiological Psychology
JCP   Journal of Comparative Psychology
JEP   Journal of Experimental Psychology
JGP   Journal of Genetic Psychology
JN   Journal of Nutrition
GPM   Genetic Psychology Monographs
PR   Psychological Review

Some of the titles of papers have been shortened. *The numbers in brackets after each entry are those of the pages on which it is mentioned.*

1 Abercrombie M. 1947 (personal communication). (98)

2 Adolph E. F. 1947 *AJP* **151**, 110-125. Urges to eat and drink in rats. (45)

3 Adolph E. F., Barker J. P., Hoy P. J. 1954 *AJP* **178**, 538-562. Multiple factors in thirst. (57)

4 Agar W. E., Drummond F. H., Tiegs O. W. 1948 *J. exp. Biol.* **25**, 103-122. Third report on a test of McDougall's Lamarckian experiment on the training of rats. (127)

5 Agar W. E., Drummond F. H., Tiegs O. W., Gunson M. M. 1954 *J. exp. Biol.* **31**, 308-321. Fourth (final) report on a test of McDougall's Lamarckian experiment on the training of rats. (127)

6 Anand B. K., Brobeck J. R. 1951 *Yale J. Biol. Med.* **24**, 123-140. Hypothalamic control of food intake in rats and cats. (68)

7 Anderson J. W. 1954 *Science* **119**, 808-809. The production of ultrasonic sounds by laboratory rats and other mammals. (160)

8 Anderson O. D., Liddell H. S. 1935 *Arch. Neurol. Psychiat.* **34**, 330-354. Observations on experimental neurosis in sheep. (199)

9 Andrew R. J. 1956 *BJAB* **4**, 41-45. Some remarks on behaviour in conflict situations. (195)

10 Anthony R. 1928 *Leçons sur le Cerveau.* Paris: Doin. (217)

11 Anthony W. S. 1959 *Brit. J. Psychol.* **50**, 117-124. The Tolman and Honzik insight situation. (149)

12 Armour C. J., Barnett S. A. 1950 *J. Hyg. Camb.* **48**, 158-170. The action of dicoumarol on laboratory and wild rats. (50)

13 Armstrong E. A. 1950 *Symp. Soc. exp. Biol.* **4**, 361-384. The nature and functions of displacement activities. (196)

14 Aschkenasy-Lelu P. 1946 *J. Psychol. norm. path.* **39**, 445-466. Le choix des aliments protéiques à fonction du besoin azoté chez le rat. (45)

15 Aschkenasy-Lelu P. 1947 *Ann. Nutrit. Alim.* **3**, 277-314. La valeur du choix spontané des aliments. (46)

16 Bagnall-Oakley R. P. 1955 *Trans. Norfolk Norw. Nat. Soc.* **18**, 17-23. (134)

17 Barnett S. A. 1955 *Lancet* **269**, 1203-1208. 'Displacement' behaviour and 'psychosomatic' disorder. (196)

18 Barnett S. A. 1956 *Behaviour* **9**, 24-43. Behaviour components in the feeding of wild and laboratory rats. (17, 36, 38, 40, 53, 194)

19 Barnett S. A. 1958 *Proc. zool. Soc. Lond.* **130**, 107-152. An analysis of social behaviour in wild rats. (41, 73, 75, 76, 77, 83, 85, 86, 88, 194, 195, 197, 202)

20 Barnett S. A. 1958 *Brit. J. Psychol.* **49**, 195-201. Experiments on 'neophobia' in wild and laboratory rats. (30)

21 Barnett S. A. 1958 *Brit. J. Psychol.* **49**, 289-310. Exploratory behaviour. (16, 164, 185)

22 Barnett S. A. 1958 *J. Psychosom. Res.* **3**, 1-11. Physiological effects of 'social stress' in wild rats. 1. The adrenal cortex. (203)

23 Barnett S. A. 1958 *J. anim. Tech. Assoc.* **9**, 6-14. Laboratory methods for the study of wild rat behaviour. (104, 121)

24 Barnett S. A. 1960 *Proc. zool. Soc. Lond.* **134**, 611-621. Social behaviour among tame rats and among wild-white hybrids. (94)

25 Barnett S. A. 1961 *Lancet* **280**, 1067-1071. The behaviour and needs of infant mammals. (208)

26 Barnett S. A. Unpublished observations. (2, 39, 41, 75, 103, 122, 191)

27 Barnett S. A., Bathard A. H. 1953 *J. Hyg. Camb.* **51**, 483-491. Population dynamics of sewer rats. (83)

28 Barnett S. A., Bathard A. H., Spencer M. M. 1951 *Ann. app. Biol.* **38**, 444-463. Rat populations and control in two English villages. (48, 83)

29 Barnett S. A., Eaton J. C., McCallum H. M. 1960 *J. Psychosom.*

Res. **4,** 251-260. Physiological effects of 'social stress' in wild rats. 2. Liver glycogen and blood glucose. (83, 202, 242)

30 Barnett S. A., Spencer M. M. 1949 *J. Hyg. Camb.* **47,** 426-430. Sodium fluoracetate (1080) as a rat poison. (49)

31 Barnett S. A., Spencer M. M. 1951 *Behaviour* **3,** 229-242. Feeding, social behaviour and interspecific competition in wild rats. (37, 41)

32 Barnett S. A., Spencer M. M. 1953 *BJAB* **1,** 32-37. Responses of wild rats to offensive smells and tastes. (35, 47, 53)

33 Barnett S. A., Spencer M. M. 1953 *J. Hyg. Camb.* **51,** 16-34. Experiments on the food preferences of wild rats. (52)

34 Barrett-Hamilton A. H., Hinton M. A. C. 1912-20 *History of British Mammals.* London: Gurney and Jackson. (2)

35 Beach F. A. 1940 *JCP* **29,** 193-245. Effects of cortical lesions on the copulatory behavior of male rats. (227, 228)

36 Beach F. A. 1944 *J. exp. Zool.* **97,** 249-295. Relative effects of androgen upon the mating behavior of male rats subjected to forebrain injury or castration. (215, 228)

37 Beach F. A. 1948 *Hormones and Behavior.* New York: Hoeber. (109)

38 Beach F. A. 1950 *Amer. Psychol.* **5,** 115-124. The Snark was a Boojum. (1)

39 Beach F. A., Hebb D. O., Morgan C. T., Nissen H. W. 1960 *The Neuropsychology of Lashley.* New York: McGraw-Hill. (223)

40 Beach F. A., Holz-Tucker A. M. 1949 *JCPP* **42,** 433-453. Mating behavior in male rats castrated at various ages and injected with androgen. (114)

41 Beach F. A., Jaynes J. 1956 *J. Mammal.* **37,** 177-180. Studies of maternal retrieving in rats. 1. Recognition of young. (91)

42 Beach F. A., Jaynes J. 1956 *Behaviour* **10,** 104-125. Studies in maternal retrieving in rats. 3. Sensory cues. (120)

43 Beach F. A., Levinson G. 1950 *J. exp. Zool.* **114,** 159-172. Effects of androgen on the glans penis and mating behavior of castrated male rats. (117)

44 Benedict F. G. 1938 *Vital Energetics.* Washington. (76)

45 Bergen J. R. 1951 *AJP* **164,** 16-22. Rat electrocorticogram in relation to adrenal cortical function. (222)

46 Berlyne D. E. 1960 *Conflict, Arousal and Curiosity.* New York: McGraw-Hill. (185, 232)

47 Bindra D. 1948 *JCPP* **41,** 397-402. What makes rats hoard? (42)

48 Bindra D. 1959 *Motivation: A Systematic Reinterpretation.* New York: Ronald Press. (160, 176)

49 Bingham W. E., Griffiths W. J. 1952 *JCPP* **45**, 307-312. The effect of different environments during infancy on adult behavior in the rat. (184)

50 Birch H. G. 1956 *Am. J. Orthopsychiat.* **26**, 279-284. Sources of order in the maternal behavior of mammals. (121)

51 Birch H. G., Korn S. J. 1958 *JGP* **58**, 17-35. Place-learning, cognitive maps, and parsimony. (165)

52 Blodgett H. C. 1929 *Univ. Calif. Publ. Psychol.* **4**, 113-134. The effect of the introduction of reward upon the maze performance of rats. (163, 164)

53 Bonin G. v. 1950 *Essay on the Cerebral Cortex.* Springfield, Illinois: Thomas. (257)

54 Broadhurst P. L. 1959 *Acta Psychol.* **16**, 321-338. The interaction of task difficulty and motivation. (170)

55 Broadhurst P. L. 1960 In Eysenck H. J. (ed.) *Handbook of Abnormal Psychology.* London: Pitman. Abnormal animal behaviour. (198)

56 Brobeck J. R. 1948 *Yale J. Biol. Med.* **20**, 545-552. Food intake as a mechanism of temperature regulation. (62)

57 Brobeck J. R. 1955 *Ann. N.Y. Acad. Sci.* **63**, 44-55. Neural regulation of food intake. (64, 69)

58 Brobeck J. R. 1957 *Yale J. Biol. Med.* **29**, 565-574. Neural control of hunger, appetite and satiety. (62)

59 Bruce H. M., Kennedy G. C. 1951 *Proc. Roy. Soc. B* **138**, 528-544. The central nervous control of food and water intake. (68)

60 Brunswick E. 1939 *JEP* **25**, 175-197. Probability as a determiner of rat behavior. (167)

61 Butler R. A. 1957 *JCPP* **50**, 177-179. The effect of deprivation of visual incentives on visual exploration motivation in monkeys. (20)

62 Buytendijk F. J. J. 1931 *Arch. Néerl. Physiol.* **16**, 574-596. Eine Methode zur Beobachtung von Ratten in aufgabefreien Situationen. (93)

63 Calhoun J. B. 1948 *J. Wild. Manag.* **12**, 167-171. Mortality and movement of brown rats (*Rattus norvegicus*) in artificially super-saturated populations. (82)

64 Calhoun J. B. 1949 *Science* **109**, 333-335. A method for self control of population growth among mammals living in the wild. (41, 54, 82)

64a Campbell B. A. 1955 *JCPP* **48**, 141-148. Reduction in stimulation to produce learning. (176)

65 Campbell B. A., Kraeling D. 1953 *JEP* **45**, 97-101. Response strength as a function of drive level and amount of drive reduction. (176)

66 Carlson A. J., Hoelzel F. 1949 *Science* **109**, 63-64. Influence of texture of food on its acceptance by rats. (51)

67 Chitty D. 1954 *The Control of Rats and Mice: vols. 1 & 2, Rats.* Oxford: Clarendon Press. (29, 36, 47, 49)

68 Christian J. J. 1956 *Ecology* **37**, 258-273. Adrenal and reproductive responses in mice from freely growing populations. (203)

69 Clark W. E. le Gros 1948 *Lancet* **254**, 353-356. The connexions of the frontal lobes of the brain. (221)

70 Clarke J. R. 1956 *Behaviour* **9**, 1-23. The aggressive behaviour of the vole. (194)

71 Conger J. J., Sawrey W. L., Turrell E. S. 1958 *J. abn. soc. Psychol.* **57**, 214-220. Social experience in the production of gastric ulcers in hooded rats. (197)

72 Cook S. W. 1939 *Am. J. Psychiat.* **95**, 1259-1276. A survey of methods used to produce 'experimental neurosis'. (199)

73 Cott H. B. 1940 *Adaptive Coloration in Animals.* London: Methuen. (87)

74 Cottam C. 1948 *J. Mammal.* **29**, 299. Aquatic habits of the Norway rat. (185)

75 Cowles J. T., Pennington L. A. 1943 *J. Psychol.* **15**, 41-47. An improved conditioning technique for determining auditory acuity of the rat. (137)

76 Crawford M. P. 1939 *Psychol. Bull.* **36**, 407-446. The social psychology of the vertebrates. (54)

77 Crespi L. P. 1942 *Amer. J. Psychol.* **55**, 467-517. Quantitative variation of incentive and performance in the white rat. (170)

77a Crozier W. J. 1928 *JGP* **1**, 213-238. Tropisms. (32)

78 Curti M. W. 1935 *Psychol. Monog.* **46**, 78-98. Native fear responses of white rats in the presence of cats. (107)

79 Darchen R. 1955 *Z. Tierpsychol.* **12**, 1-11. Stimuli nouveaux et tendance exploratrice chez *Blattella germanica*. (28)

80 Darchen R. 1957 *J. Psychol. norm. path.* **54**, 190-205. Sur le comportement d'exploration de *Blattella germanica*. Exploration d'un plan. (28)

81 Darwin C. 1859 *On the Origin of Species.* London: Murray. (127)

82 Darwin C. 1873 *On the Expression of the Emotions in Man and Animals.* London: Murray. (127, 191)

83 Davis D. E. 1951 *Ecology* **32,** 459-461. The relation between level of population and pregnancy of Norway rats. (81)

84 Davis D. E., Emlen J. T., Stokes A. W. 1948 *J. Mammal.* **29,** 207-225. Studies on home range in the brown rat. (15)

85 Deutsch J. A., Jones A. D. 1959 *Nature* **183,** 1472. The water-salt receptor and preference in the rat. (58)

86 Dodwell P. C., Bessant D. E. 1960 *JCPP* **53,** 422-425. Learning without swimming in a water maze. (162)

87 Donaldson H. H. 1924 *The Rat: Data and Reference Tables.* Philadelphia: H. H. Donaldson. (3)

88 Dove C. C., Thompson M. E. 1943 *JGP* **63,** 235-245. Some studies on 'insight' in white rats. (149)

89 Eayrs J. T., Moulton D. G. 1960 *Quart. J. exp. Psychol.* **7,** 90-98. Studies in olfactory acuity. I. Measurement of olfactory thresholds in the rat. (35)

90 Eccles J. C. 1953 *The Neurophysiological Basis of Mind.* Oxford: Clarendon Press. (113, 220)

91 Eibl-Eibesfeldt I. 1955 *Naturwiss.* **23,** 633-634. Angeborenes und Erworbenes in Nestbauverhalten der Wanderratte. (121)

92 Emlen J. T., Stokes A. W., Winser C. P. 1948 *Ecology* **29,** 133-145. The rate of recovery of decimated populations of brown rats in nature. (82)

93 Epstein A. N., Stellar E. 1955 *JCPP* **48,** 167-172. The control of salt preference in the adrenalectomized rat. (66)

94 Farber I. E. 1948 *JEP* **38,** 111-131. Response fixation under anxiety and non-anxiety conditions. (178)

95 Finan J. L. 1940 *JCP* **29,** 119-134. Quantitative studies in motivation. I. Strength of conditioning in rats under varying degrees of hunger. (171)

96 Finley C. B. 1941 *J. comp. Neurol.* **54,** 203-237. Equivalent losses in accuracy of response after central and after peripheral sense deprivation. (225)

97 Forgays D. G., Forgays J. W. 1952 *JCPP* **45,** 322-328. The nature of the effect of free-environmental experience in the rat. (184)

98 Forgus R. H. 1954 *JCPP* **47,** 331-336. The effect of early perceptual learning on the behavioral organization of adult rats. (185)

99 Forgus R. H. 1955 *JCPP* **48**, 215-220. Early visual and motor experience as determiners of complex maze-learning ability. (184, 185)

100 Forgus R. H. 1956 *Canad. J. Psychol.* **10**, 147-155. Advantage of early over late perceptual experience in improving form discrimination. (185)

101 Forgus R. H. 1958 *JCPP* **51**, 588-591. The interaction between form pre-exposure and test requirements in determining form discrimination. (185)

102 French J. D. 1958 In [162]. Corticifugal connections with the reticular formation. (236)

103 French J. D., Porter R. W., Cavanaugh E. B., Longmire R. L. 1954 *Arch. Neurol. Psychiat.* **72**, 267-281. Experimental observations on 'psychosomatic' mechanisms. 1. Gastrointestinal disturbances. (196)

104 French J. D., Porter R. W., Cavanaugh E. B., Longmire R. L. 1957 *Psychosom. Med.* **19**, 209-220. Experimental gastroduodenal lesions induced by stimulation of the brain. (196)

105 French J. D., Sugar O., Chusid S. G. 1948 *J. Neurophysiol.* **11**, 185-192. Corticocortical connections of the superior bank of the sylvian fissure in the monkey (*Macacca mulatta*). (226)

106 Fuller J. L., Thompson W. R. 1960 *Behavior Genetics*. New York: Wiley. (257)

107 Fulton J. F. 1949 *Functional Localization in the Frontal Lobes and Cerebellum*. Oxford: Clarendon Press. (225)

108 Funkenstein D. 1955 *Sci. Am.* **92** (v), 74-80. The physiology of fear and anger. (202)

109 Gastaut H. 1958 In [162]. The role of the reticular formation in establishing conditioned reactions. (235)

110 Gelineo S., Gelineo A. 1952 *Bull. Acad. serbe Sci. math. nat.* **4**, 197-210. La température du nid du rat et sa signification biologique. (75, 105)

111 Gellhorn E. 1953 *Physiological Foundations of Neurology and Psychiatry*. Minneapolis: University of Minnesota Press. (201, 203)

112 Gellhorn E. 1957 *Autonomic Imbalance and the Hypothalamus*. Minneapolis: University of Minnesota Press. (201, 202)

113 Gibson E. J., Walk R. D. 1960 *Sci. Am.* **202** (iv), 64-71. The 'visual cliff'. (108)

114 Glickman S. E. 1958 *Canad. J. Psychol.* **12**, 45-51. Effects of peripheral blindness on exploratory behaviour in the hooded rat. (25)

115 Griffin D. R. 1958 *Listening in the Dark*. New Haven: Yale University Press. (160)

116 Griffith C. R. 1920 *Psychobiol.* **2,** 19-28. The behavior of white rats in the presence of cats. (107)

117 Grossman M. I. 1955 *Ann. N.Y. Acad. Sci.* **63,** 76-89. Integration of current views on the regulation of hunger and appetite. (59)

118 Haldane J. B. S. 1954 *Behaviour* **6,** 256-270. A logical analysis of learning, conditioning and related processes. (142)

119 Haldane J. B. S. 1954 In *L'instinct dans le comportement des animaux et de l'homme*. Paris: Singer-Polignac. Les aspects physico-chimiques des instincts. (192)

120 Haldane J. B. S. 1956 *BJAB* **4,** 162-164. The sources of some ethological notions. (192)

121 Haldane J. B. S., Spurway H. 1956 *Nature* **178,** 85-86. Imprinting and the evolution of instincts. (128, 129)

122 Harlow H. F. 1932 *JGP* **41,** 211-220. Social facilitation of feeding in the albino rat. (55)

123 Harlow H. F. 1949 *PR* **56,** 51-65. The formation of learning sets. (182)

124 Harlow H. F. 1959 *Amer. Sci.* **47,** 459-479. The development of learning in the rhesus monkey. (182)

125 Harlow H. F., Woolsey C. N. 1958 *Biological and Biochemical Bases of Behavior*. Madison: University of Wisconsin Press. (156, 229, 231, 232)

126 Harris L. J., Clay J., Hargreaves F. J. and Ward A. 1933 *Proc. Roy. Soc. B* **113,** 161-190. Appetite and choice of diet. (45)

127 Harris V. T. 1952 *Contr. Lab. Vert. Biol. Mich. No. 56*. An experimental study of habitat selection by prairie and forest races of the deermouse *Peromyscus maniculatus*. (128)

128 Harte R. A., Travers J. J., Sarich P. 1948 *JN* **36,** 667-680. Voluntary caloric intake of the growing rat. (44)

129 Hausmann M. F. 1933 *JCP* **15,** 419-428. The behavior of albino rats in choosing foods. (44)

130 Hebb D. O. 1937 *JGP* **51,** 101-126. The innate organization of visual activity. 1. Perception of figures by rats reared in total darkness. (183)

131 Hebb D. O. 1946 *PR* **53,** 259-276. On the nature of fear. (30, 242)

132 Hebb D. O. 1947 *Psychosom. Med.* **9,** 3-19. Spontaneous neurosis in chimpanzees. (198)

133 Hebb D. O. 1949 *The Organization of Behavior*. New York: Wiley. (63, 153, 155, 179, 183, 214, 218, 220)

134 Hebb D. O. 1955 *PR* **62**, 243-254. Drives and the C.N.S. (conceptual nervous system). (63, 221)

135 Hebb D. O. 1956 *Canad. J. Psychol.* **10**, 165-166. The distinction between 'classical' and 'instrumental'. (146)

136 Hebb D. O. 1958 *A Textbook of Psychology*. Philadelphia: Saunders. (132)

137 Hebb D. O., Mahut H. 1955 *J. Psychol. norm. path.* **52**, 209-221. Motivation et recherche du changement perceptif chez le rat et chez l'homme. (27)

138 Hebb D. O., Williams K. 1946 *JGP* **34**, 59-65. A method of rating animal intelligence. (153)

139 Hediger H. 1950 *Wild Animals in Captivity*. London: Butterworth. (24, 78, 97)

140 Hermann G. 1958 *Z. Tierpsychol.* **15**, 462-518. Beiträge zur Physiologie des Rattenauges. (3)

141 Hernández-Peón R. 1955 *J. Latinoam.* **1**, 256. Central mechanisms controlling conduction along central sensory pathways. (264)

142 Herrick C. J. 1924 *An Introduction to Neurology*. Philadelphia: Saunders. (35)

143 Hess E. H. 1959 *Science* **130**, 133-141. Imprinting. (207)

144 Hill W. F. 1958 *JCPP* **51**, 570-574. The effect of varying periods of confinement on activity in tilting cages. (20)

145 Hinde R. A. 1954 *BJAB* **2**, 41-55. Changes in responsiveness to a constant stimulus. (102)

146 Hinde R. A. 1954 *Proc. Roy. Soc. B* **142**, 306-331. Factors governing the changes in strength of a partially inborn response. (143)

147 Hinde R. A. 1956 *Ibis* **98**, 340-369. The territories of birds. (92)

148 Hinde R. A. 1956 *Brit. J. Philos. Sci.* **6**, 321-331. Ethological models and the concept of 'drive'. (190)

149 Hinde R. A. 1958 *Proc. zool. Soc. Lond.* **131**, 1-48. The nest-building behaviour of domesticated canaries. (104)

150 Hinde R. A. 1959 *AB* **7**, 130-141. Unitary drives. (189)

151 Hinde R. A. 1959 *Biol. Rev.* **34**, 85-128. Behaviour and speciation in lower vertebrates. (97, 113)

152 Hinde R. A. 1960 *Symp. Soc. exp. Biol.* **14**, 199-213. Energy models of motivation. (191)

153 Hinde R. A., Thorpe W. H., Vince M. A. 1956 *Behaviour* **9,** 214-242. The following response of young coots and moorhens. (207)

154 Hogben L. T. 1957 *Statistical Theory*. London: Allen & Unwin. (13)

155 Holst E. von 1954 *BJAB* **2,** 89-94. Relations between the central nervous system and the peripheral organs. (214)

156 Hull C. L. 1935 *PR* **42,** 219-245. The mechanism of the assembly of behavior segments in novel combinations. (149, 161)

157 Hull C. L. 1943 *Principles of Behavior*. New York: Appleton-Century. (161)

158 Hull C. L. 1952 *A Behavior System*. New Haven: Yale University Press. (172, 204)

159 Hydén H. 1960 In Brachet J. & Mirsky A. E. (ed.) *The Cell*. **4,** 215-323. The neuron. (237)

160 Iersel J. J. A. van, Bol A. C. A. 1958 *Behaviour* **13,** 1-88. Preening of two tern species. A study on displacement activities. (196)

161 Janowitz H., Grossman M. I. 1948 *AJP* **155,** 28-32. Effects of parenteral administration of glucose and protein hydrolysate on food intake in the rat. (61)

162 Jasper H. H., Proctor L. D., Knighton R. S., Noshay W. C., Costello R. T. 1958 *Reticular Formation of the Brain*. London: Churchill. (233)

163 Jerome E. A., Moody J. A., Connor T. J., Fernandez M. B. 1957 *JCPP* **50,** 588-591. Learning in multiple-door situations under various drive states. (19)

164 Kagan J., Beach F. A. 1953 *JCPP* **46,** 204-211. Effects of early experience on mating behavior in male rats. (119)

165 Kagan J., Berkun M. 1954 *JCPP* **47,** 108. The reward value of running activity. (25)

166 Kalmus H. 1955 *BJAB* **3,** 25-31. The discrimination by the nose of the dog of individual human odours. (91)

167 Karli P. 1956 *Behaviour* **10,** 81-103. The Norway rat's killing response to the white mouse. (94)

168 Kendler H. H. 1946 *JEP* **36,** 212-220. The influence of simultaneous hunger and thirst drives upon the learning of two opposed spatial responses of the white rat. (205)

169 Kennedy G. C. 1950 *Proc. Roy. Soc. B* **137,** 535-549. The hypothalamic control of food intake in rats. (68, 192)

170 Kennedy G. C. 1953 *Proc. Roy. Soc. B* **140**, 578-596. The role of depot fat in the hypothalamic control of food intake in the rat. (62)

171 Kennedy G. C. 1961 *Proc. Nutr. Soc.* **20**, 58-64. The central nervous regulation of calorie balance. (62)

172 Kinder E. F. 1927 *J. exp. Zool.* **47**, 117-161. A study of the nest-building activity of the albino rat. (104)

173 King J. A. 1958 *Psychol. Bull.* **55**, 46-58. Parameters relevant to determining the effect of early experience upon the adult behavior of animals. (182)

174 Kish G. B., Antonitis J. J. 1956 *JGP* **88**, 121-129. Unconditioned operant behavior in two homozygous strains of mice. (25)

175 Kohn M. 1951 *JCPP* **44**, 412-421. Satiation of hunger from food injected directly into the stomach, versus food injected by mouth. (61)

176 Koronakos C., Arnold W. J. 1957 *JCPP* **50**, 11-14. The formation of learning sets in rats. (183)

177 Krech D., Rosenzweig M. R., Bennett E. L. 1960 *Science* **132**, 352-353. Interhemispheric effects of cortical lesions on brain biochemistry. (237)

178 Krech D., Rosenzweig M. R., Bennett E. L. 1960 *JCPP* **53**, 509-519. Effects of environmental complexity and training on brain chemistry. (237)

179 Krechevsky I. 1932 *PR* **39**, 516-532. Hypotheses in rats. (150)

180 Krechevsky I. 1933 *JCP* **16**, 99-116. Hereditary nature of 'hypotheses'. (124)

181 Krechevsky I. 1937 *JCP* **23**, 139-159. Brain mechanisms and variability II. Variability where no learning is involved. (227)

182 Lack D. 1954 *The Natural Regulation of Animal Numbers.* Oxford: Clarendon. (81, 92)

183 Lansdell H. C. 1953 *JCPP* **46**, 461-464. Effect of brain damage on intelligence in rats. (225)

184 Larsson K. 1956 *Acta Psychol. Gothobergensis I.* Conditioning and sexual behaviour in the male rat. (109, 119)

185 Lashley K. S. 1929 *Brain Mechanisms and Intelligence.* Chicago: University of Chicago Press. (155, 224)

186 Lashley K. S. 1934 In Murchison C. (ed.) *Handbook of General Experimental Psychology.* Worcester: Clark University Press. Nervous mechanisms in learning. (108, 161)

187 Lashley K. S. 1938 *JGP* **18**, 123-193. The mechanism of vision. XV. Preliminary studies of the rat's capacity for detail vision. (216)

188 Lashley K. S. 1949 *Quart. Rev. Biol.* **24**, 28-42. Persistent problems in the evolution of mind. (213)

189 Lashley K. S. 1950 *Symp. Soc. exp. Biol.* **4**, 454-482. In search of the engram. (223)

190 Lashley K. S., Russell J. T. 1934 *JGP* **45**, 136-144. The mechanism of vision. XI. A preliminary test of innate organization. (107)

191 Lát J. 1956 *Physiol. Bohemoslov.* **5** (suppl.), 38-42. The relationship of the individual differences in the regulation of food intake, growth and excitability of the central nervous system. (80, 124)

192 Lát J., Widdowson E. M., McCance R. A. 1960 *Proc. Roy. Soc. B* **153**, 347-356. Some effects of accelerating growth. 3. Behaviour and nervous activity. (80, 209)

193 Lehrman D. S. 1955 *Behaviour* **7**, 241-286. The physiological basis of parental feeding behaviour in the ring dove (*Streptopelia risoria*). (117)

194 Lehrman D. S. 1956 In Grassé P.-P. et al. *L'instinct dans le Comportement des Animaux et de l'Homme*. Paris: Fondation Singer-Polignac. On the organisation of maternal behaviour and the problem of instinct. (116, 119)

195 Lepkovsky S. 1948 *Adv. Food Res.* **1**, 105-148. The physiological basis of voluntary food intake. (46)

196 Levine S. 1960 *Sci. Am.* **202** (v), 81-86. Stimulation in infancy. (208)

197 Levine S., Soliday S. 1960 *JCPP* **53**, 497-501. The effects of hypothalamic lesions on conditioned avoidance learning. (203)

198 Liddell H. S. 1942 In Stone C. S. (ed.) *Comparative Psychology*. New York: Prentice-Hall. The conditioned reflex. (138)

199 Liddell H. S. 1956 *Emotional Hazards in Animals and Man*. Springfield, Illinois: Thomas. (199)

200 Lissmann H. W. 1950 *Symp. Soc. exp. Biol.* **4**, 34-59. Proprioceptors. (213)

201 McCleary R. A. 1954 *JCPP* **47**, 411-421. Taste and post-ingestion factors in specific-hunger behavior. (58)

202 McCulloch W. S. 1944 In Bucy P. C. (ed.) *The Precentral Motor Cortex*. Urbana: University of Illinois Press. Cortico-cortical connections. (226)

203 McCulloch W. S. 1950 *Comp. Psychol. Monog.* **20,** 39-50. Machines that think and want. (214)

204 McDougall W. 1938 *Brit. J. Psychol.* **28,** 321-345, 365-395. Fourth report on a Lamarckian experiment. (126)

205 Magnen J. le 1951 *CRSB* **145,** 851-854, 854-857, 857-860. Étude des phénomènes olfacto-sexuels chez le rat blanc. (78, 111, 116).

206 Magoun H. W. 1958 *The Waking Brain.* Springfield, Illinois: Thomas. (233, 235)

207 Maier N. R. F. 1949 *Frustration.* New York: McGraw-Hill. (200)

208 Maier N. R. F., Schneirla T. C. 1935 *Principles of Animal Psychology.* New York: McGraw-Hill. (148, 162)

209 Marler P. 1959 In Bell P. R. (ed.) *Darwin's Biological Work.* Cambridge: University Press. Developments in the study of animal communication. (79)

210 Marx M. H. 1950 *PR* **57,** 80-93. A stimulus-response analysis of the hoarding habit in the rat. (43)

211 Marx M. H. 1957 *JCPP* **50,** 168-171. Experimental analysis of the hoarding habit in the rat. III. Terminal reinforcement under low drive. (42)

212 Marx M. H., Brownstein A. J. 1957 *JCPP* **50,** 617-620. Experimental analysis of the hoarding habit in the rat. IV. Terminal reinforcement followed by high drive at test. (42)

213 Masserman J. H. 1946 *Principles of Dynamic Psychiatry.* Philadelphia. (199, 201)

214 Michie D. 1956 *Lab. Anim. Bur. coll. Pap.* **3,** 37-47. Towards uniformity in experimental animals. (3)

215 Miller G. A., Viek P. 1944 *JCP* **37,** 221-231. An analysis of the rat's response to unfamiliar aspects of the hoarding situation. (42)

216 Miller N. E. 1944 In Hunt J. M. (ed.) *Personality and the Behavior Disorders.* New York: Ronald Press. Experimental studies in conflict. (199)

217 Miller N. E. 1948 *JEP* **38,** 89-101. Studies of fear as an acquirable drive. (177)

218 Miller N. E. 1955 *Ann. N.Y. Acad. Sci.* **63,** 141-143. Shortcomings of food consumption as a measure of hunger. (58, 64)

219 Miller N. E., Dollard J. 1945 *Social Learning and Imitation.* London: Routledge. (55)

220 Miller N. E., Sampliner R. I., Woodrow P. 1957 *JCPP* **50,** 1-5. Thirst-reducing effects of water by stomach fistula versus water by mouth. (58)

221 Mirsky I. A., Miller R., Stein M. 1953 *Psychosom. Med.* **15,** 574-584. Relation of adrenocortical activity and adaptive behavior. (203)

221a Montgomery K. C. 1951 *JCPP* **44,** 582-589. Relation between exploratory behavior and spontaneous alternation. (21)

222 Montgomery, K. C. 1953 *JCPP* **46,** 315-319. The effect of hunger and thirst drives upon exploratory behavior. (19)

223 Montgomery K. C. 1953 *JCPP* **46,** 438-441. The effect of activity deprivation on exploratory behavior. (20)

224 Montgomery K. C., Zimbardo P. G. 1957 *Percept. Mot. Skills* **7,** 223-229. Effect of sensory and behavioral deprivation upon exploratory behavior in the rat. (25)

225 Morgan C. T. 1947 *PR* **54,** 335-341. The hoarding instinct. (42, 43)

226 Morgan C. T., Stellar E., Johnson O. 1943 *JCP* **35,** 275-295. Food deprivation and hoarding in rats. (41)

227 Morrison S. D., Mayer J. 1957 *AJP* **191,** 248-254. Adipsia and aphagia in rats after lateral subthalamic lesions. (69)

228 Morrison S. D., Mayer J. 1957 *AJP* **191,** 255-258. Effect of sham operations in the hypothalamus on food and water intake of the rat. (69)

229 Moulton D. G., Eayrs J. T. 1960 *Quart. J. exp. Psychol.* **7,** 99-109. Studies in olfactory acuity. II. Relative detectability of *n*-aliphatic alcohols by the rat. (35)

230 Mowrer O. H. 1960 *Learning Theory and Behavior.* New York: Wiley. (144, 163, 176)

231 Mowrer O. H. 1960 *Learning Theory and the Symbolic Process.* New York: Wiley. (37, 140, 163, 180)

232 Mowrer O. H., Jones H. M. 1945 *JEP* **35,** 293-311. Habit strength as a function of the pattern of reinforcement. (172)

233 Munn N. L. 1950 *Handbook of Psychological Research on the Rat.* New York: Houghton Mifflin. (41, 55, 59, 73, 74, 80, 109, 126, 136, 144, 152, 207)

234 Neuhaus W. 1950 Schädlingsbekämpfung **42,** 108-111. (35)

235 O'Kelly L. I., Heyer A. W. 1951 *Comp. Psychol. Monog.* **20,** 287-298. The influence of need duration on retention of a maze habit. (169, 171)

236 Olds J. in [125]. Adaptive functions of paleocortical and related structures. (175, 176)

237 Olds J. 1959 *Ann. Rev. Physiol.* **21**, 381-402. High functions of the nervous system. (204)

238 Orgain H., Schein M. W. 1953 *Ecology* **34**, 467-473. A preliminary analysis of the physical environment of the Norway rat. (83)

239 Osgood C. E. 1953 *Method and Theory in Experimental Psychology.* New York: Oxford University Press. (163, 166, 181, 223, 225)

240 Patrick J. R., Laughlin R. M. 1934 *JGP* **44**, 378-389. Is the wall-seeking tendency in the white rat an instinct ? (32)

241 Peretz E. 1960 *JCPP* **53**, 540-548. The effects of lesions of the anterior cingulate cortex on the behavior of the rat. (230)

242 Petrinovich L., Bolles R. 1957 *JCPP* **50**, 363-365. Delayed alternation. (153)

243 Pfaffmann C. 1957 *Amer. J. Clin. Nutr.* **5**, 142-147. Taste mechanisms in preference behavior. (66)

244 Pisano R. G., Storer T. I. 1948 *J. Mammal.* **29**, 374-383. Burrows and feeding of the Norway rat. (2)

245 Prokasy W. F. 1956 *JCPP* **49**, 131-134. The acquisition of observing responses in the absence of differential external reinforcement. (180)

246 Prosser C. L., Hunter W. S. 1936 *AJP* **117**, 609-618. The extinction of startle responses and spinal reflexes in the white rat. (141, 142)

247 Prychodko W. 1958 *Ecology* **39**, 500-503. Effect of aggregation of laboratory mice (*Mus musculus*) on food intake at different temperatures. (75)

248 Quigley J. P. 1955 *Ann. N.Y. Acad. Sci.* **63**, 6-14. The role of the digestive tract in regulating the ingestion of food. (63)

249 Rabinovitch M. S., Rosvold H. E. 1951 *Canad. J. Psychol.* **5**, 122-128. A closed-field intelligence test for rats. (153, 154)

250 Reese W. G., Dykman R. A. 1960 *Physiol. Rev.* **40**, 250-265. Conditional cardiovascular reflexes in dogs and men. (137)

251 Reiff M. 1952 *Verh. Schweiz. Naturf. Ges. Luzern 1951*, 150-151. Uber Territorieumsmarkierung bei Hausratten und Hausmausen. (78)

252 Reiff M. 1953 *Rev. Suisse Zool.* **60**, 447-452. Differenzierungen im ökologischen Verhalten bei Wanderrattenpopulationen. (78)

253 Reyniers J. A., Ervin R. F. 1946 Lobund Reports (i) 1-84. (80)

254 Richter C. P. 1937 *Cold Spr. Harb. Symp. Quant. Biol.* **5,** 258-268. Hypophysial control of behavior. (104)

255 Richter C. P. 1942 *Harvey Lect.* **38,** 63-103. Total self-regulatory functions in animals and human beings. (17, 46, 115)

256 Riess B. F. 1950 *Ann. N.Y. Acad. Sci.* **51,** 1093-1102. The isolation of factors of learning and native behavior in field and laboratory studies. (121)

257 Riley D. A., Rosenzweig M. R. 1957 *JCPP* **50,** 323-328. Echolocation in rats. (160).

258 Ritchie B. F. 1947 *JEP* **37,** 25-38. Studies in spatial learning. (160)

259 Robinson R. 1950 *Definition.* Oxford: Clarendon Press. (247)

260 Roe A., Simpson G. G. (ed.) 1958 *Behavior and Evolution.* New Haven: Yale University Press. (156)

261 Rose J. E., Woolsey C. N. 1949 *J. comp. Neurol.* **91,** 441-466. The relations of thalamic connections, cellular structure and evocable electrical activity in the auditory region of the cat. (218)

262 Rosenblatt J. S., Aronson L. R. 1958 *AB* **6,** 171-182. The influence of experience on the behavioural effects of androgen in prepuberally castrated male cats. (117)

263 Rosenzweig M. R., Krech D., Bennett E. L. 1960 *Psychol. Bull.* **57,** 476-492. A search for relations between brain chemistry and behavior. (237)

264 Russell B. 1927 *An Outline of Philosophy.* London: Allen & Unwin. (241)

265 Russell R. W. 1953 *Eug. Rev.* **45,** 19-30. Experimental studies of hereditary influences on behaviour. (124)

266 Russell R. W., Pretty R. G. F. 1951 *Quart. J. exp. Psychol.* **3,** 151-156. A study of position habits induced by reward and 'frustration'. (200)

267 Rzoska J. 1953 *BJAB* **1,** 128-135. Bait shyness, a study in rat behaviour. (47, 48, 51)

268 Rzoska J. 1954 In Chitty D. (ed.) *The Control of Rats and Mice, vol. 2.* Oxford: Clarendon Press. The behaviour of white rats towards poison baits. (47, 51)

269 Schein M. W., Orgain H. 1953 *Amer. J. Trop. Med. Hyg.* **2,** 1117-1130. A preliminary analysis of garbage as food for the Norway rat. (82)

270 Schneirla T. C. 1959 In Jones M. R. (ed.) *Nebraska Symposium on Motivation*. Lincoln: University of Nebraska Press. An evolutionary and developmental theory of biphasic processes underlying approach and withdrawal. (162)

271 Schorstein J. 1955 *Surgo* 131-142. Experiences with closed head injuries in a neurosurgical unit 1948-54. (243)

272 Scott E. M. 1948 *Trans. Amer. Assoc. Cereal Chem.* **6,** 126-133. Self selection of diet. (46)

273 Scott E. M., Quint E. 1946 *JN* **32,** 113-120. Self selection of diet. II. The effect of flavor. (53)

274 Scott E. M., Quint E. 1946 *JN* **32,** 285-292. Self selection of diet. III. Appetite for B vitamins. (46)

275 Scott E. M., Quint E. 1946 *JN* **32,** 293-302. Self selection of diet. IV. Appetite for protein. (45)

276 Scott E. M., Verney E. L. 1948 *JN* **36,** 91-98. Self selection of diet. VIII. Appetite for fats. (52)

277 Scott E. M., Verney E. L. 1949 *JN* **37,** 81-92. Self selection of diet. IX. The appetite for thiamine. (46)

278 Scott E. M., Verney E. L., Morissey P. D. 1950 *JN* **41,** 173-186. Self selection of diet. X. Appetites for sodium, chloride and sodium chloride. (46)

279 Scott E. M., Verney E. L., Morissey P. D. 1950 *JN* **41,** 187-202. Self selection of diet. XI. Appetites for calcium, magnesium and potassium. (46)

280 Searle L. V. 1949 *GPM* **39,** 279-325. The organization of hereditary maze-brightness and maze-dullness. (124)

281 Seitz P. F. D. 1954 *Am. J. Psychiat.* **110,** 916-927. The effects of infantile experiences upon adult behavior in animal subjects. 1. Effects of litter size during infancy upon adult behavior in the rat. (80)

282 Sellers E. A., Reichman S., Thomas N. 1951 *AJP* **167,** 644-650. Acclimatization to cold. (45)

283 Sellers E. A., Reichman S., You S. S. 1951 *AJP* **167,** 651-655. Acclimatization to cold in rats: metabolic rates. (45)

284 Seward J. P. 1945 *JCP* **38,** 213-224. Aggressive behavior in the rat. 2. An attempt to establish a dominance hierarchy. (93)

285 Sharpless S., Jasper H. H. 1956 *Brain* **79,** 655-680. Habituation of the arousal reaction. (235)

286 Sheffield F. D., Roby T. B. 1950 *JCPP* **43,** 471-481. Reward value of non-nutritive sweet taste. (53)

287 Sheffield F. D., Roby T. B., Campbell B. A. 1954 *JCPP* **47,**

349-354. Drive reduction versus consummatory behavior as determinants of reinforcement. (53)

288 Sheffield F. D., Wulff J. J., Backer R. 1951 *JCPP* **44**, 3-8. Reward value of copulation without sex drive reduction. (174)

289 Sherrington C. 1952 *The Integrative Action of the Nervous System*. Cambridge: University Press. (212)

290 Skinner B. F. 1938 *The Behavior of Organisms*. New York: Appleton-Century. (26, 145, 171)

291 Small W. S. 1899 *Amer. J. Psychol.* **11**, 80-100. Notes on the psychic development of the young white rat. (16, 93, 155)

292 Small W. S. 1900 *Amer. J. Psychol.* **11**, 133-165. An experimental study of the mental processes of the rat. (93, 144)

293 Smith M., Duffy M. 1957 *JCPP* **50**, 601-608. Some physiological factors that regulate eating behavior. (60, 61)

293a Smith M. P., Capretta P. J. 1956 *JCPP* **49**, 553-557. Effects of drive level and experience on the reward value of saccharin solutions. (53)

294 Solomon R. L. 1948 *Psychol. Bull.* **45**, 1-40. The influence of work on behavior. (160)

295 Soulairac A. 1952 In *Structure et Physiologie des Sociétés Animales*. Paris: Centre National de la Recherche Scientifique. L'effet de groupe dans le comportement sexuel du rat mâle. Étude expérimentale du comportement en groupe du rat blanc. (74, 93)

296 Soulairac A. 1958 *J. Physiol. Path. gen.* **50**, 663-783. Les régulations psycho-physiologiques de la faim. (63)

297 Soulairac A., Soulairac M. L. 1954 *CRSB* **148**, 304-307. Effets du groupement sur le comportement alimentaire du rat. (55)

298 Spence K. W. 1956 *Behavior Theory and Conditioning*. New Haven: Yale University Press. (169, 205)

299 Spencer M. M. 1953 *J. Hyg. Camb.* **51**, 35-38. The behaviour of rats and mice feeding on whole grains. (40)

300 Spurway H., Haldane J. B. S. 1953 *Behaviour* **6**, 8-34. The comparative ethology of vertebrate breathing. 1. Breathing in newts, with a general survey. (102)

301 Stamm J. S. 1953 *JCPP* **45**, 299-304. Effects of cortical lesions on established hoarding activity in rats. (229)

302 Steinberg H., Watson R. H. J. 1960 *Nature* **185**, 615-616. Failure of growth in disturbed laboratory rats. (94)

303 Stone C. P. 1929 *GPM* **5**, 1. The age factor in animal learning. I. Rats in the problem box and maze. (207)

304 Stone C. P.   1929   *GPM*   **6**, 2. The age factor in animal learning. II. Rats in a multiple light discrimination box. (207)

305 Stone C. P., Darrow C. W., Landis C., Heath L. L.   1932   In *Studies in the Dynamics of Behaviour*. Chicago. The heredity of wildness and savageness in rats. (123)

306 Strominger J. L., Brobeck J. R.   1953   *Yale J. Biol. Med.*   **25**, 383-390. A mechanism of regulation of food intake. (67)

307 Teitelbaum P.   1957   *JCPP*   **50**, 486-490. Random and food-directed activity in hyperphagic and normal rats. (192)

308 Teitelbaum P., Stellar E.   1954   *Science*   **120**, 894-895. Recovery from the failure to eat produced by hypothalamic lesions. (69)

309 Thompson H. V.   1948   *Bull. Anim. Behav.*   **6**, 26-40. Behaviour of the common brown rat taking plain and poisoned bait. (39)

309a Thompson W. R.   1960   *Am. J. Orthopsychiat.*   **30**, 306-314. Early environmental influences on behavioral development. (123)

310 Thompson W. R., Higgins W. H.   1958   *Canad. J. Psychol.* **12**, 61-68. Emotion and organized behaviour. (195)

311 Thorpe W. H.   1951   *Bull. Anim. Behav.*   **9**, 34-40. The definition of terms used in animal behaviour studies. (7, 134, 252)

312 Thorpe W. H.   1956   *Learning and Instinct in Animals*. London: Methuen. (28, 109, 130, 134, 135, 142, 143, 162, 179)

313 Thorpe W. H., Zangwill O. L. (ed.)   1961   *Current Problems in Animal Behaviour*. Cambridge: University Press. (218, 223)

314 Tinbergen N.   1952   *Quart. Rev. Biol.*   **27**, 1-32. "Derived" activities. (111, 194)

315 Tinbergen N.   1953   *Social Behaviour in Animals*. London: Methuen. (111)

316 Tolman E. C.   1948   *PR*   **55**, 189-208. Cognitive maps in rats and men. (162)

317 Tolman E. C.   1958   *Behavior and Psychological Man*. Berkeley: University of California Press. (93, 144)

318 Tolman E. C., Honzik C. H.   1930   *Univ. Calif. Publ. Psychol.* **4**, 215-232. "Insight" in rats. (148, 168)

319 Tribe D. E., Gordon J. G.   1953   *Brit. J. Nutrit.*   **7**, 197-201. Choice of diet by rats deficient in members of the vitamin B complex. (46)

320 Tryon R. C.   1940   *Yearb. Nat. Soc. Stud. Educ.*   **39**, 111-119. Genetic differences in maze-learning ability in rats. (124)

321 Tsang Y.-C. 1934 *Comp. Psychol. Monog.* **10**, 1-56. The function of the visual areas of the cortex of the rat in the learning and retention of the maze. (143, 225)

322 Valenstein E. S. V., Riss W. R., Young W. C. 1955 *JCPP* **48**, 397-403. Experimental and genetic factors in the organization of sexual behavior in the male guinea pig. (119)

323 Verplanck W. S. 1957 *PR (Suppl.)* **64** (ii). A glossary of some terms used in the objective science of behavior. (147, 249)

324 Viek P., Miller G. A. 1944 *JCP* **37**, 203-210. The cage as a factor in hoarding. (42)

325 Vincent S. B. 1913 *J. comp. Neurol.* **23**, 1-36. The tactile hair of the white rat. (31)

326 Waddington C. H. 1956 *Evolution* **10**, 1-13. Genetic assimilation of the *bithorax* phenotype. (128)

326a Wang G. H. 1923 *Comp. Psychol. Monog.* **2** (6), 1-27. Relation between 'spontaneous' activity and oestrus cycle in the white rat. (115)

327 Warner L. H. 1932 *JGP* **41**, 57-90. The association span of the white rat. (145)

328 Weinstock S. 1954 *JCPP* **47**, 318-322. Resistance to extinction of a running response after partial reinforcement. (171)

329 Weiss B. 1957 *JCPP* **50**, 481-485. Thermal behavior of the subnourished and pantothenic-acid-deprived rat. (105)

330 Wiesner B. P., Sheard N. M. 1933 *Maternal Behaviour in the Rat*. Edinburgh: Oliver & Boyd. (79, 80, 105, 120)

331 Wodinsky J., Bitterman M. E. 1953 *Amer. J. Psychol.* **66**, 137-140. The solution of oddity problems by the rat. (153)

332 Wolff H. G. 1953 *Stress and Disease*. Springfield, Illinois: Thomas. (196)

333 Wolpe J. 1958 *Psychotherapy by Reciprocal Inhibition*. Stanford: University Press. (199)

334 Wolsey C. N. 1952 In *The Biology of Mental Health and Disease*. London: Cassell. Patterns of localization in sensory and motor areas of the cerebral cortex. (219)

335 Wynne L. C., Solomon R. L. 1955 *GPM* **52**, 241-284. Traumatic avoidance learning in dogs deprived of normal peripheral autonomic function. (177)

336 Young J. Z. 1951 *Proc. Roy. Soc. B* **139**, 18-37. Growth and plasticity in the nervous system. (220)

337 Young J. Z. 1957 *The Life of Mammals*. Oxford: Clarendon Press. (257)

338 Young J. Z. 1961 *Biol. Rev.* **36,** 32-96. Learning and discrimination in the octopus. (161, 212, 214)

339 Young P. T. 1949 *PR* **56,** 98-121. Food-seeking drive, affective process, and learning. (46)

340 Young P. T., Shuford E. H. 1955 *JCPP* **48,** 114-118. Quantitative control of motivation through sucrose solutions of different concentrations. (169)

341 Young P. T., Wittenborn J. R. 1940 *JCP* **30,** 261-276. Food preference of rachitic and normal rats. (46)

342 Young W. C. 1957 In Hoagland H. (ed.) *Hormones, Brain Function and Behaviour.* New York: Academic Press. Genetic and psychological determinants of sexual behavior patterns. (110, 119)

343 Zimbardo P. G., Montgomery K. C. 1957 *Psychol. Rep.* **3,** 589-594. Effects of 'free-environment' rearing upon exploratory behavior. (27)

344 Zuckerman S. 1932 *The Social Life of Monkeys and Apes.* London: Kegan Paul. (84)

Index